THE PATH TO MORE SUSTAINABLE ENERGY SYSTEMS

THE PATH TO MORE SUSTAINABLE ENERGY SYSTEMS

HOW DO WE GET THERE FROM HERE?

BEN W. EBENHACK AND DANIEL M. MARTÍNEZ

MOMENTUM PRESS

MOMENTUM PRESS, LLC, NEW YORK

The Path to More Sustainable Energy Systems: How Do We Get There from Here?

First published by Momentum Press®, LLC
222 East 46th Street, New York, NY 10017
www.momentumpress.net

ISBN-13: 978-1-60650-260-0 (hardback)
ISBN-10: 1-60650-260-3 (hardback)
ISBN-13: 978-1-60650-262-4 (e-book)
ISBN-10: 1-60650-262-X (e-book)

DOI: 10.5643/9781606502624

Cover design by Jonathan Pennell
Cover image by Daniel M. Martínez
Interior design by Exeter Premedia Services Private Ltd.
Chennai, India

10 9 8 7 6 5 4 3 2 1

Printed in the United States of America

Contents

PREFACE

ENERGY USE AND TRANSITIONS

The world stands at the brink of sweeping global energy transitions—and they should be transitions toward greater sustainability. This demands attention to both sides of sustainability: meeting the needs of people today, while preserving opportunity for the future. This book is meant to shed some light on what can be known about the energy options, their potential, and their limitations.

What will the next transitions be like—and what should they be like? What do we want from energy? Which options can best meet those needs, and in what time frames? Currently, much of the world needs more energy. Fossil fuels provide more than 80% of the world's energy, but those supplies do face ultimate limits. A set of environmental impacts is clearly observable with the enormity of energy they offer. Can we maximize the benefits derived from energy consumption, while minimizing the costs? Which paths offer the most promise? There is not enough information to know precisely when energy shortages will impose transitions. There is no clear consensus of which energy sources or production and conversion systems will dominate the new energy landscape. There are large questions about how to compare the merits and limitations of various systems. However, there are some things that we can know. Energy is vital to survival and development. The Developing World will almost certainly demand more energy to support development—as well as growing population levels. The prevailing fossil fuel systems will be called upon heavily to meet these needs for many decades to come. They provided tremendous benefit for societies as they moved from raw biomass dependence to more modern energy systems and that transition will continue in much of the Developing World, due to the proven efficacy and cost-effectiveness of fossil fuels, even as the world embarks on new paths to new, nondepleting, and (ideally) less polluting energy systems.

To understand the transition, the first four chapters of the book build foundational information, with Chapter 1 presenting descriptions of the concepts of sustainability and energy systems, along with means to measure energy, work, and power. Since the resources commonly dubbed "nonrenewable" so dominate the world's energy supply, Chapter 2 is devoted to discussing the occurrence and acquisition of these resources. Oil and gas receive the most detailed attention because they govern such a large share of the marketplace and their shortages will propel the need for transitions. Coal and fissile material for nuclear power provide resource stocks that can be tapped, if their reserves are developed alongside necessary conversion and end-use technologies.

Generally, the "renewables" are considered to be sustainable, yet they each have their own constraints. Chapter 3 discusses the range of renewable resources, noting that biomass has much in common with fossil fuels in terms of emissions and the potential for humans to deplete stocks.

Even the processed biofuels impact other resource systems. Although solar and wind energy have enormous resource bases, which cannot be depleted, they have only finite fluxes to tap and currently produce a tiny fraction of humanity's energy. Large-scale, dam-based hydropower provides more energy than the other sustainable alternatives, but its future development is probably limited by concerns over ecosystem disruption. Small-scale and run-of-the-river technologies may represent some of the potential for the future of hydropower, along with forays into wave and tidal power. Finally, geothermal power is a resource that we suggest deserves more attention than it receives.

Energy is produced from primary energy sources and transformed into energy carriers and through end-use technologies to provide useful services to consumers. Whereas the energy itself is harnessed, rather than being truly consumed, Chapter 4 takes up the essential task of describing and characterizing the processes by which energy is provided to consumers. Sectoral analyses help to trace how energy resources are utilized and to what purposes. Different sectors demand more energy as development proceeds. Humans first used external energy to provide heat and light. Even in modern times, cooking remains the dominant energy demand for people in lower income countries. This demand is still met by raw firewood, charcoal, or even dung. Process heat is needed once the most basic levels of manufacturing are introduced. More intense processes need superior fuels. For about one million years, our ancestors' use of fire did not change significantly. They gathered and burned biomass directly. There is some evidence that some societies declined as they consumed more firewood than was locally available—or they were forced to find ways to bring in more firewood or to innovate new technologies. Indeed, it was shortages of firewood in England that forced experimentation with coal. Knowledge about energy consumption patterns promotes an understanding of energy in the contexts of whole systems, from sources, to conversions to end use—and waste products.

The first of two general types of constraints on energy would be the physical limits of production, taken up and analyzed in some detail in Chapter 5. The nature of exponential growth is important to understand in recognizing the inevitability of limits. The concept of "peak oil" is explored in some depth, as it will play such an important role in the transition. The controversy surrounding what we can or cannot know about peak oil is explored and evidence is offered to show that the peak is inevitable, probably by the middle of the 21st century, but that petroleum is likely to remain a vital part of global energy production for several decades (and plausibly another century) after the peak. This creates a time horizon for transitions within which other resources will need to develop. The constraints on each system are discussed.

The second type of constraint involves sustainability assessments that take into account the environmental impacts of each option. Chapter 6 endeavors to characterize these impacts. In recent years, there has been increasing focus on the environmental costs of our energy sources—primarily the fossil fuels: air pollution, climate change, and oil spills. This focus causes us to lose sight of the benefits that we receive from our energy resources. The costs must not be ignored, but neither must the values. We must find ways to work constructively toward prudent use of the resources that we have and toward a conscientious transition to the resources of the future. Environmental impacts are discussed in many texts (and sometimes presented as if sustainability is solely an environmental issue); this chapter seeks to summarize much of what is understood about the environmental issues and offer perspective on the breadth of impacts that can be attributed to each energy system.

Since sustainability demands meeting human needs and caring for the future, it is best thought of as a problem of optimizing the values provided to humanity over time, considering resource availability, technologic capacity, and negative impacts. Chapter 7 seeks to bring the various costs and benefits together. We offer a means to evaluate the correlation between energy and quality of life. The evidence suggests a very strong correlation. It exhibits a saturation-type behavior, with a little more energy consumption correlating to huge gains for energy poor people, but diminishing gains with ever more consumption, until a limiting level is reached at which more energy seems to generate no additional gains. This strongly suggests the need for the Developing World to consume more energy.

So we find ourselves in the beginning of the 21st century with residents of affluent, industrialized nations consuming exponentially growing amounts of energy. The commercial energy markets are dominated by petroleum and many of us are wondering how long petroleum will be able to meet the growing appetites of a growing global population. When will there be an energy shortage? For half of humanity, the answer is simple. It's already here. For the affluent Developed World, the question is how long our appetites may be sated. For the people in rapidly developing countries, the question is what energy systems can support their growth both now and into the future, while minimizing environmental costs. The complexity of optimizing disparate sorts of costs and benefits is discussed, with some recommendations about how efforts may proceed to do this in a transparent and reasonably robust fashion.

Finally, in Chapter 8 all of the information considered is brought together to evaluate how best we can begin to plan a pathway toward more sustainable energy futures. The challenge will be a new "voyage of discovery" more akin to the exploration of Lewis and Clark than to the subsequent pioneers who followed known paths to settle the frontiers. There is no one clear path to follow. Energy issues are markedly different in the Developed World than in the Developing World. No one solution is likely to be adequate even within a single context, but surely not across such a large divide. The resources, the consumption patterns, the levels of need, the infrastructural capacities, and the opportunities are all radically different. We suggest that it will be important to plan energy transitions that are contextually appropriate. The problem for the Developed World is largely how to maintain the benefits we currently derive from our energy consumption as we approach limits to critical resources. The problem for the other half of humanity is how to gain the benefits of adequate, modern energy in a world competing for the control of the resources. We can define some general directions, based on the goals and the limits laid out in earlier chapters as well as the contexts of place and time. Perhaps most vitally, we can identify some paths that are counterproductive.

Changes will be immense. The Developing World needs more energy, while the consumption levels in the Developed World will be challenged by impending petroleum shortages, even in affluent, energy-rich and disproportionately consumptive countries like the United States and Canada. It is unlikely that any of the popular "renewable energy" sources will be able to rise to the challenge in the requisite time frame. Nuclear fission has the potential to provide a great deal more energy, but will it be a viable alternative in the Developing World, where the need is the greatest? The world's largest consumer (the United States) has tremendous potential for conservation, but what will be required to effect that conservation without sacrificing quality of life? How far into the future is nuclear fusion's great promise? What resources will go into developing the alternative energy systems and new infrastructure to support them? How do issues of public perception, societal needs, technical maturity, and scale inform or affect the path toward more sustainable energy systems for the world?

Without access to reliable electricity, any and all development efforts are severely hampered. It seems rather unsurprising to us that the hundreds of billions of dollars of international aid spent over the last few decades have failed to lift the people in developing countries out of poverty, considering the utter disregard for essential energy services. Around 1990, the ambassador to the United States from one developing country reported a conversation with a senior international aid official.

The Ambassador accosted the official to urge support of an energy development project.
That official replied, "No, what you need is agricultural aid."
The ambassador replied, "You give us tractors and we have no fuel to run them."

This exchange poignantly exemplifies the critical fallacy of ignoring energy development. It fails to see energy as a critical part of other systems. Without energy, truly, no modern development can be supported.

We hope that this book offers engineering professionals, policy makers, students, and the public alike, a useful text for which to consider the concept of energy sustainability in the 21st century. We also hope that our knowledge and our experiences in energy evaluation and project planning both in the Developed and in the Developing World come through in our writing to offer a balanced, perhaps even sober view of the challenges and opportunities that await us. Due to the extraordinary breadth of the energy systems topic, we have drawn on the work of many others as well and urge readers to look to our extensive citations to guide one toward greater details. We have deliberately included many website citations, as they may be updated more frequently in the rapidly changing landscape of global energy systems.

We would like to thank our wives, Mary Jeanette and Thalia, both for assisting and for tolerating us in this protracted endeavor. We would also like to thank the students and colleagues who volunteered (or were coerced) to review early drafts of this text and offer input. Additionally, we would like to thank those at Momentum Press for their help and patience during the process of producing this manuscript, especially as deadlines were extended to accommodate rewrites, new jobs, the rearing of a young boy, and the birth of a beautiful baby girl. Any errors or oversights found in this text are no fault of anyone save the authors.

Finally, we would like to end, or rather begin, with an old Irish anecdote of a fellow asking directions to Dublin. The reply is, "Well, I wouldn't start from here." We cannot realistically talk about where we are going on any journey without understanding from whence we begin. There will still be many unknowns, though, as we embark on the discovery of new and more sustainable territory.

Ben Ebenhack
Marietta College
Marietta, Ohio, USA

Daniel Martínez
University of Southern Maine
Gorham, Maine, USA

Key Words/Terms

sustainable energy; sustainable development; energy; sustainability; development; environmental sustainability; energy sustainability; peak oil; energy transitions; environmental; sustainable; fossil fuels; renewable energy; energy resources; energy access; energy supply; Developing World; oil & gas; environmental science; pollution; energy systems

CHAPTER 1

CONCEPTS, DEFINITIONS, MEASURES

This chapter introduces basic energy concepts and provides background information for the rest of the book.

1.1 DEFINING ENERGY

Energy is such a fundamental aspect of nature that it permeates all that we perceive. All that our eyes see is light energy either generated by a source or reflected. Sound is the result of atoms set in motion by energy. Even all of the physical objects that surround us can be considered to be energy manifested as mass or matter. Energy can exist either in flux or manifested in its interactions with matter. We most commonly observe energy in flux as electromagnetic radiation (e.g., light and heat).

In its material manifestations, energy can exist in kinetic, potential, or nuclear forms. Kinetic energy is the energy of matter in motion, such as a moving car or wind. Potential energy normally refers to gravity, in which energy is stored by moving mass away from the dominant gravitational center—it can be released by allowing the mass to fall toward the center of gravity (for our purposes, Earth). Chemical energy is another important form of potential energy stored in the bonds of molecules. It can be released by breaking those bonds and recombining them into less energetic forms. Nuclear energy taps into the form of energy that exists as mass. Only a tiny amount of mass is destroyed while releasing tremendous quantities of nuclear energy.

Some specific concepts and their units of measure that are important to energy use are highlighted below.

1.1.1 WORK

Work is defined as "a measure of the amount of 'change' that a force produces when it acts on a body" (Beiser 2009). Mathematically, this can be expressed as a body's mass, multiplied by its acceleration, multiplied by the length over which it changed. Since acceleration can be defined as a length divided by time squared, we can further simplify work to be dependent on only three basic measureable variables. In the SI or metric system, a unit of work is expressed specifically as a kilogram (mass) times a meter (length) squared divided by a second (time) squared ($kg \cdot m^2/s^2$), also known as a joule (J), whereas in the United States Customary System (USCS) a unit of work is known as a foot-pound ($ft \cdot lb$). Work and energy are interchangeable concepts, as energy simply is a measureable quantity of work—and the more energy something has, the more work it can do.

1.1.2 HEAT

Heat, also known as thermal energy, is the movement and vibration of the pieces that make up a body or substance. Thus, when added to a body, heat will increase its internal energy, in turn causing its temperature to rise. Because heat is a form of energy, the SI unit of heat is the joule, however, it is also common to use the calorie (cal), which is the amount of heat required to raise the temperature of one gram of water by one degree Celsius. In the USCS, the unit of heat is the British thermal unit (BTU), which is the amount of heat required to raise the temperature of one pound of water by one degree Fahrenheit. (The Calories we count in our meals are actually kilocalories—thousands of the calories used in physics.)

1.1.3 LIGHT

Radiant energy is the energy of light: including both visible and nonvisible portions of the spectrum. For this energy to be seen, "there must be a source to provide the radiant energy and a receptor to receive and translate this energy to an image" (Reist 1993). A source, such as the sun, emits radiant energy, which can be measured in joules. In the study of visible light, a parallel definition has emerged, with the source of luminous energy being measured in lumen-seconds ($lm \cdot s$).

1.1.4 ELECTRICITY

Electrical energy is the result of electrons flowing across a voltage (potential) difference. The flow creates an electric field or current, whose measure is work divided by charge. The unit of potential difference is the volt (V), which is equal to a joule divided by a coulomb.

1.1.5 POWER

Power is the rate at which work (or energy) is done. Mathematically, this can be expressed as work divided by a time interval. In the SI, a unit of power is expressed specifically as a joule divided by second (J/s), also known as a watt (W), whereas in the USCS a unit of power is known as a foot-pound divided by a second ($ft \cdot lb/s$), also known as a horsepower (hp). In the study of heat, power is also referred to as a heat rate (W) and a heat flux equal to the heat rate per unit area (W/m^2). In the study of radiant energy, power is commonly measured as a radiant flux, emittance, or irradiance (W/m^2). In the study of visible light, luminous flux is measured in lumens (lm), and luminous emittance or illuminance measured in lux (lm/m^2). In the study of electricity, power is typically reported in watts, which is the rate at which work is done to hold an electric current, equal to a potential difference multiplied by a current. As Beiser (2009) also points out, under certain circumstances, power also can be expressed as a force multiplied by a velocity. Power is reflected in acceleration in common observations and it has the same units as energy production: energy per unit time. Similar to energy, the more power something has, the more work it can do in a given time interval.

1.1.6 EFFICIENCY

The efficiency of a system is simply the ratio of the energy (or power) the system takes in from a source and the energy (or power) the system then transmits away from it. That is:

$$\text{Efficiency} = (\text{output})/(\text{input})$$

Essentially, efficiency describes how well a system can convert an energetic input into some useful output. Although it seems desirable to value only the systems with high efficiencies of conversion, one should keep in mind that some of the most important energy conversion systems on earth have extremely small efficiency values (e.g., photosynthesis).

1.2 KEY ENERGY RESOURCE DEFINITIONS

The nature of energy itself may seem rather esoteric. In this book, we are mostly concerned with energy sources and resources, and how they might be used more sustainably by humans.

1.2.1 SOURCES AND RESOURCES

An energy source refers to that from which energy originates or can be tapped for human use. It should be distinguished from an energy carrier, such as electricity or hydrogen, which requires more energy input to generate than it contains. Of course it is important to quantify energy sources and be able to assess their potential contributions to global energy needs. The global resource base for any energy source, then, would be the total amount believed to exist in the world. It can also be defined within any particular geographical boundaries, but carries no limitations either of the economy or of proof by exploration. For petroleum, the resource base is also referred to as the "total petroleum initially-in-place," which would include quantities already produced. The resource base is assessed by educated, but still fairly wild guesses.

1.2.2 RESERVES

Reserves, on the other hand, represent the amount of an energy source believed to be recoverable under prevailing economic and technological constraints. Indeed, proved reserves are only those that have already been discovered and begun to produce. The proved reserves are a subset of reserves, which are a relatively small subset of the resource base. Figure 1.1 presents the most current terminology used by the Society of Petroleum Engineers (SPE) to characterize quantities of fossil fuels, based on the level of certainty that they can be produced. Production is subtracted from reserves, while new discoveries or enhanced recovery potential are added to reserves.

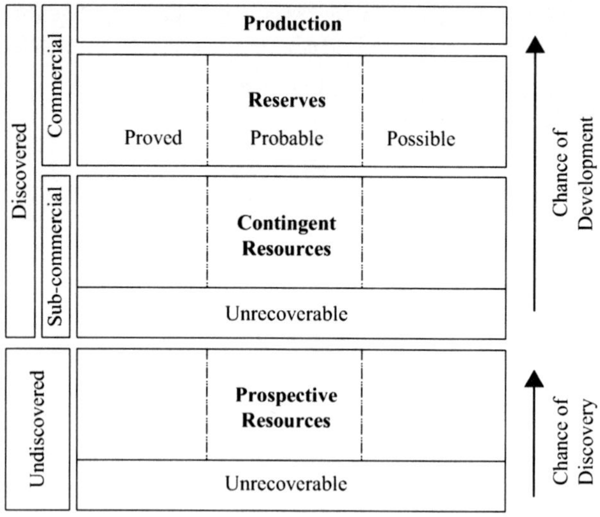

Figure 1.1. Quantity characterization of fossil fuels.
Source: SPE (2010).

1.2.3 PRODUCTION

Production is the amount of energy that has been or is currently being derived from the energy source. (For solar, wind, wave, and run-of-river hydropower, there are no reserves, only production.) The top left of the table depicted in Figure 1.1 represents the absolute quantities of production (the production is measured directly). Moving to the right on the table represents less certainty of the resource being commercially recoverable, based either on decreasing strength of data or on the need for new technology that has not yet been utilized in a given field. Moving down the table refers to decreasing confidence in the commerciality of production. Currently identified, but subcommercial resources are still more certain than those which have not yet been discovered. Note, as well that, without production, there can be no proved reserves for an energy source.

It is important to keep the general classifications in mind when comparing energy sources to alternatives. The reason is: it is tempting to compare the resource base of one energy system to the reserves of another, but because the resource base is inevitably several times larger than the reserves, this is clearly an inconsistency. Moreover, it lends to the tendency to commit the error of comparing the hypothetical potential of an alternative that has little or no established production to the proved recoverable reserves of a current energy mainstay. If no production has ever been established commercially, then its reserves must be zero. We will address this issue a little later.

1.2.4 COMPARING UNITS AND MAGNITUDES OF MEASURE

Another difficulty in comparing energy sources is the wide array of units used to describe them. Oil reserves are presented in terms of barrels (or millions or billions of barrels.) A barrel of oil consists of 42 gallons, with a variable energy content averaging approximately 5.8 million BTUs per barrel. (Recall that a BTU is the amount of energy required to raise 1 pound of water

1°F.) Oil production and consumption rates are typically expressed in some factor of barrels of oil per day (BOPD). Quantities of natural gas are shown in terms of thousands of cubic feet (MCF.) Note that the M used in gas measurement is the Roman numeral representing 1000, not an abbreviation for million. Gas is a somewhat more uniform commodity than crude oil, with an energy content consistently around 1000 BTUs per cubic foot or 1 million BTUs per MCF. In the case of a "wet gas," the presence of substantial liquid content in the gas raises the BTU value at the wellhead, but the liquids are commonly stripped out and bottled as condensate before the gas goes to market, leaving the final product with an energy content very close to 1000 BTU per cubic foot.

Some of the most common units for energy are:

1 BTU equals 1055 J (joules) equals 252 calories equals 0.000293 kWh (kilowatt hours or kilowatts times the number of hours they are being produced)

Since the world is now tapping vast energy flows, numerous prefixes are employed to represent appropriate orders of magnitude:

1 MJ (megajoule) equals 1 million or 10^6 J
1 EJ (exajoule) equals 10^{18} J
1 Quad equals one quadrillion or 10^{15} BTUs equals 1.055 EJ

In spite of the variable energy content of coal and oil, they are dominant energy sources and are often used as references for other energy sources:

1 tce (ton of coal equivalent) equals 27.7 8 million BTUs
1 toe (ton of oil equivalent) equals 40 million BTUs
1 boe (barrel of oil equivalent) equals 5.8 million BTUs

Note that these conversions are not precise, because even the standard units may have variable definitions. For instance, there is a 0.1% range in the energy content represented by one BTU, based on pressure and temperature conditions at which it is measured. These minor variations will have no noticeable effect on the assessments in this book of resource bases for predictions about production or depletion, since there are much greater uncertainties involved in all of these analyses than the minor variation of definitions for energy conversion units.

1.3 "RENEWABLE" VERSUS "NONRENEWABLE" ENERGY

Energy sources are commonly divided into two categories: "renewable" and "nonrenewable." However, we see problems with this convention for a number of reasons. One important reason is that many resources tend to straddle both categories under various conditions and really do not lend to being defined in absolutes. Another important reason has to do with the use of "renewability" as a category in general, because it incorrectly (although not purposely) connotes that resources falling into this category are somehow limitless in their utilization. Indeed, both nonrenewable and renewable resources are limited—the former by stocks and the latter by flows. In her book, *Thinking in Systems*, Donnella Meadows (2008) makes an excellent effort to describe these concepts, which we will both try to summarize and expand on below.

1.3.1 STOCK AND FLOW LIMITATIONS

Nonrenewable resources are generally considered to be the fossil fuels and uranium. The fossil fuels represent millions of years of accumulated sun-earth capital. Uranium is an ancient endowment of fissile material resulting from primordial supernovae. Our use of these finite resources is almost entirely dictated by the amount of mineral deposits known to be available for exploitation, and the economic conditions that allow for their extraction from the ground. This means, for example, that the entire nonrenewable mineral stock on earth is, theoretically, "available at once" (Meadows 2008), with identified reserves having the chance to be added to in the form of improved recovery via technology enhancements or via new discoveries. Reserve stocks also can be increased by decreasing the extraction rate (in consumption sectors) via increased efficiency of conversion, or via conservation. Ultimately though, the resource base—representing the total stock of the resource—is not renewed (at least not on useful timeframes) and the faster it is extracted for use, the shorter the lifetime of that resource. *Thus, nonrenewable resources are stock-limited resources.*

Renewable resources are those resources that can be extracted in near "real time" and can be harvested indefinitely (again, at least on useful timeframes). But, as Meadows puts it, this can be done "only at a finite flow rate equal to their regeneration rate." She further states that if the rate of extraction occurs faster than the rate of regeneration, "they may eventually be driven below a critical threshold and become, for all practical purposes, nonrenewable." Implicit in this definition is the point that when using renewable resources, there are no substantial stocks from which to draw, except in certain instances, and even that is available for very short periods of time compared with the accumulated stocks available from the nonrenewable resources, which took eons to form. *Thus, renewable resources are flow-limited resources.*

1.3.2 FOSSIL AND NUCLEAR FUELS: NONRENEWABLE, STOCK-LIMITED ENERGY

When we think of examples of nonrenewable energy, we tend to think of the fossil fuels: coal, petroleum, and gas. As Vaclav Smil (2008) put it, "fossil fuels were formed through slow but profound changes of accumulated biomass under pressure and heat, and their ages range from millions to hundreds of millions of years." The quantities available are immense and demonstrate the gargantuan potential of solar energy when it is concentrated over many, many years.

Finally, we add that technically speaking, the fossil fuels are being renewed in nature right now. However, it is likely that several civilizations will rise and fall before any humans surviving those events could benefit from the exploitation of these new deposits. What is unsettling about this statement, and as Smil (2008) correctly points out, is that this accumulated solar capital is being drawn down by humans "at rates that will exhaust it in a tiny fraction of the time that was needed to create it." The nuclear fuel precursor uranium is being drawn down in a similar fashion, but this mineral is inherent to earth and its oceans and cannot be replenished.

1.3.3 SOLAR ENERGY: RENEWABLE, FLOW-LIMITED ENERGY

Classic examples of renewable energy are solar energy and anything derived from it: movement of air from wind and waves, movement of water as part of the hydrologic cycle, and biomass generated from photosynthesis. We choose only to consider solar, wind and wave, and run-of-river hydropower to be truly renewable, flow-limited energy, because for these kinds of resources, the amount used today has no direct effect on how much is available tomorrow.

For example, we consider solar energy to be renewable because the influx of solar radiation to earth is relatively constant and, more importantly, not diminished by humanity's use of it. No matter how much solar energy we use, we do nothing to deplete the sun. We only collect the radiation for heat or the light for electricity. Similarly, the present and future availability of wind, wave, and minimally manipulated hydropower cannot be affected by our current consumption of energy derived from them.

We are, however, constrained by renewable, flow-limited energy sources and their relative availability. When the sun is not shining or the wind is not blowing adequately onto their respective collectors, no substantial energy is available for utilization without the availability of some sort of integrated chemical or other storage. The intermittency of these sources makes them very difficult to depend on as standalone primary energy.

1.3.4 IN-BETWEEN RESOURCES: RENEWABLE, STOCK, AND FLOW-LIMITED ENERGY

Photosynthesis, the hydrologic cycle, and the radioactive decay that leads to geothermal heating are all continuous earth-bound processes and an annual energy flux can be plausibly estimated for these sources. However, all of these have been harnessed by humanity in a manner that is more similar to how we utilize the fossil fuels. That is, we tap accumulated, short-term stocks of these certain types of renewable energy that can and do replenish at rates much quicker than fossil fuel replenishment rates. However, there are two caveats. First, the quantity of the accumulated stocks is much smaller than the accumulated stocks of the fossil fuels. Second, it is possible to deplete the stocks being tapped to a point where the rate of depletion of these finite accumulations exceeds the flux of the process to replenish those stocks. Thus, the energy resources tapped from these inherent processes—biomass energy, stored hydropower, and geothermal energy—are stock and flow-limited energy. They possess a combination of the advantages and disadvantages of both flow and stock-based resources.

1.3.4.1 Biomass

The use of biomass for energy is an ancient practice, and the harnessing of this energy in a traditional society would have taken "a few months (crops harvested for food and fuel), a few years (... shrubs, young trees), or a few decades (mature trees) to become usable" (Smil 2008). Because photosynthesis is a continuous process, biomass is commonly referred to as renewable. Certainly, it is possible to consume biomass at rates that do not significantly diminish the existing stock, but it is possible to over-consume them as well. And while it further can be argued that it is possible for humans to contribute to the renewal of biomass (re-planting what is harvested), it is also true that most consumption of biofuels has not been at a full renewal rate, thus we have depleted reserves (e.g., forests) wherever urbanizing societies have been or currently remain dependent on biomass as a primary energy source. Modern biofuel production has some potential to be managed better in terms of re-planting, but increased biofuel production is very likely to encroach on and erode the soils of lands needed for other agricultural products or on forests and natural grasslands. In this case, the land is a finite resource.

1.3.4.2 Stored Hydropower

Stored hydropower systems that we typically associate with the hydro resource utilize dams that restrict the natural flow of water in the hydrologic cycle, thus creating a reservoir of stored gravitational energy. The larger this dam is, the larger the reservoir is, and the larger the stock of available energy that can be used for producing useful electricity. In essence, this man-made system converts a flow-limited process into a stock-limited process at specific sites, but flow limitations are perhaps a bit more important in this system because hydropower plants still depend on water flow—if water does not collect upstream of the dam, less water will flow through the hydropower plant. Again, because hydropower depends on the hydrologic cycle, it is typically considered to be renewable, however this convention has come into question because large dam systems in particular alter both the natural flow of the river and sedimentation patterns to the extent that the reservoir behind the dam ultimately silts up and may become useless. Also, tapping a natural energy flux means that it is not as available elsewhere, as it would have been if not tapped.

1.3.4.3 Geothermal Energy

Geothermal energy is particularly difficult to classify. First there are two distinct versions of geothermal energy use: small-scale, localized use for heating and cooling; and larger scale use for electric power generation. In the first case, geothermal heating and cooling clearly represent a renewable, flow-limited system. The small-scale use is unlikely to have a noticeable effect on the energy flux near the Earth's surface. Furthermore, these systems are commonly used for both heating and cooling, so heat is both extracted from and returned to the ground. Large-scale geothermal power production, though, generally involves drilling wells into geologic reservoirs that are unusually hot, then producing steam up through a well-bore, like an artificial geyser. The large-scale geothermal fields do extract tremendous amounts of heat with the steam produced and ultimately cool the reservoir. Since the steam reservoir—a stock—is surrounded by hot rock, it can deplete, but can be expected to build back up after production is halted for sufficiently long.

1.3.5 BRIEFLY COMPARING CURRENT USE OF ENERGY STOCKS AND FLOWS

According to the United States Energy Information Administration (EIA), recently, the world has been consuming roughly 400 Quads of energy per year, or 13 TW of power, from stock-limited fossil resources (EIA 2010). This represents around 85% of the total annual global energy consumption with the remaining 15% coming mostly from nuclear electric power, hydropower, and biomass, which are also mostly stock-limited energy. A very small percentage comes from flow-limited wind and solar energy. This should come as little surprise because stock-limited energy is an easier energy to utilize in modern society. As Meadows put it, stock-limited energy permits "life to proceed with some certainty, continuity, and predictability." The stores are rather sizeable and exist right now to draw from.

But, as a consumption-heavy humanity moves forward this century with the assurance that existing nonrenewable stocks will diminish substantially by around 2050 (if not sooner), it will

Table 1.1. Comparing renewable energy fluxes to current fossil energy consumption

Plausible Flux	EJ/yr	Quad/yr	TW	Stock-Limited	Stock and Flow-Limited	Flow-Limited
Runoff*	64	60	2		X	
Biomass*	390	366	12		X	
Fossil Fuel Use** (2010)	425	398	13	X		
Geothermal***	1023	959	32		X	
Wind, Waves, Currents***	11,700	10,969	366			X
Solar Radiation on Land***	765,000	717,188	24,000			X

* Data taken from Smil (2006)
** Data taken from EIA (2011)
*** Data taken from Boyle (2004)

be important to compare the potential of other sources to current consumption. Table 1.1 tabulates energy potential from a number of renewable sources, accounting for plausibly accessible (although not necessarily exploitable) fluxes and compares it to the 400 Quads per year of fossil fuel use (the current rate).

It is evident from Table 1.1 that the greatest potential comes from solar and wind resources; however, this is flow-limited energy. It does not align well with how humanity uses energy presently, and represents an immense obstacle to any proposed transition away from nonrenewable, stock-limited energy consumption.

1.4 ENERGY USE IN SOCIETIES

For much of our existence we humans have relied on nonfossilized fuels for survival and productivity. As per Hern (1999) and the *Renewable Energy Focus Handbook* (Sorensen et al. 2009), Pleistocene "societies" first drew from the quick, virtually real-time transformation of solar energy for plant-based food at a per capita energy consumption (PCEC) rate of about 3500 kilocalories per person per day, or 5.1 million BTUs (equal to roughly 37 gallons of crude oil) per person per year. Upon the discovery of fire and improved diet through cooking, this annual PCEC rate may have increased to about 8.7 million BTUs (roughly 63 gallons). Once humans moved to the slower transformations that come from deliberately planted crops for food and fuel coupled to the use and consumption of animals and the crafting of ancient building materials in agricultural societies, this annual rate likely jumped to about 23.4 million BTUs (167 gallons).

Societies reached the level of deliberately settling land near forests so as to harness the energy available in mature trees to burn biomass to heat dwellings and make charcoal for high heat applications, as well as providing for basic food needs. With this transition, the annual PCEC rate would have skyrocketed to over 160 million BTUs (1,159 gallons) per person. By this point, populations were too large to sustainably support this practice and the negative environmental effects (deforestation, soil erosion, etc.) proved that even this simple societal structure was becoming problematic to support. That is, until the use of modern fuels and revolutionary technologies became prominent.

Thanks to wide-scale mechanization coupled to abundant fossil energy, the 160 million BTU societal "barrier" was broken through and as of this writing annual global PCEC rate is hovering at about 738 million BTUs (5,347 gallons). Harkening back to Smil (2008), we know that "the affordable abundance of more efficiently-used fossil energies has transformed every productive sector of the modern economy." It is this new "normal" that the world must cope with as it grapples with massive consumption of energy in the face of a tighter supply of energy resources that can adequately interface with modern machines and infrastructures of a world population that has just surpassed seven billion people.

1.4.1 VISUALIZING ENERGY USE

There are a number of ways to visualize how we use (nonfood) energy and how it flows through our economies from primary energy source, to its conversion to fuels and electricity, to end use in residential, commercial, and industrial applications. It is informative to begin with a simple hierarchical structure of energy transfer as depicted in Figure 1.2.

At its foundation, current economies of the Industrial and Developing World depend on massive amounts of primary energy mostly from fossil fuels (coal, oil, and gas) or from biomass (wood, charcoal, and dung). The remainder comes from nuclear fuel, hydropower and "new" renewable energy (modern biofuels, wind, geothermal, and solar). Converted from this primary energy are the liquids, heat, and electricity needed for consumption choices ranging from cooking and lighting, to the heating of buildings, to the manufacture and transport of goods and services demanded by modern economies. Ultimately these choices result in a higher quality of life that then reinforces continued demand for more primary energy to sustain or

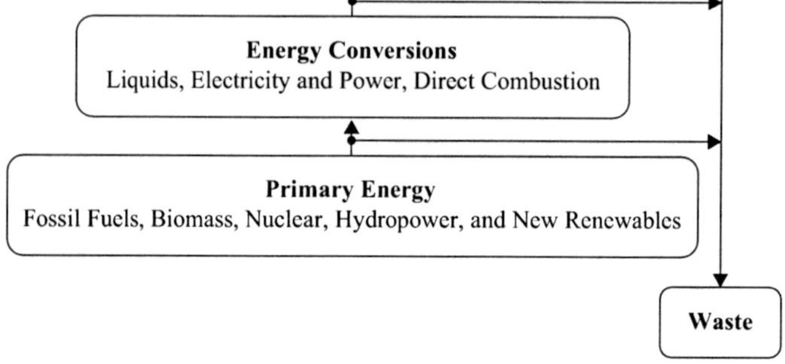

Figure 1.2. The hierarchy of energy transfer and outcomes in society.

improve on this quality, if possible. But with this demand also comes immense waste of energy and nonenergy resources.

1.4.2 ENERGY USE BY ECONOMIC SECTOR

Energy consumption data are typically reported as energy use by each economic sector, which is defined a bit differently from country to country. Following the standards of the EIA, energy consumption can be categorized into five economic sectors: (1) electric power; (2) transportation; (3) industrial; (4) residential; and (5) commercial. Often, residential and commercial sectors are lumped together, reducing this number to four. As mentioned above, the main energy sources (what we have been referring to as primary energy) used to drive these sectors are petroleum, natural gas, coal, nuclear electric power, and so-called renewable energy. Essentially, these raw fuels are processed (through a series of energy conversions that vary depending on the primary source fuel), delivered (via tankers, pipelines, and electric grids), and finally used by the consumers of these sectors.

1.4.2.1 *The Electric Power Sector*

Electricity is a convenient, robust, and extremely safe carrier of energy, and forms the basis for much economic activity in the world. It is used in heating and cooling, light and sound, communication and computation, and cooking and refrigeration. Although not commonly associated with transportation, it is also an essential component in delivering liquids and gases through the use of electric pumps. Indeed the conversion of both carbon and noncarbon-based energy to electricity has allowed for a dramatic shift in how people within social structures interact. Because of this, the electric power sector demands vast amounts of primary energy to then convert to electricity and then to transport it to other economic sectors. This conversion occurs either via combustion of carbon-based fuels (including biomass and waste) or by the energetic conversion of renewable energy.

1.4.2.2 *The Transportation Sector*

The transportation sector is primarily made up of personal vehicles, public transportation, airplanes, land and sea freight transportation, and pipelines. Worldwide energy use in this sector has grown dramatically and consistently over the past 200 years, but especially for ground and air transport (Boyle et al. 2003). Perhaps the greatest achievements in transforming how people and goods travel across land, air, and sea has happened quite recently and primarily has to do with the development and improvement of the internal combustion engine, where fuel is burned inside the engine itself. This need for liquid fuels to run the engines of transportation has resulted in, perhaps, the most one-sided dependence on a resource: petroleum.

1.4.2.3 *The Industrial Sector*

Industrialization is a relatively recent phase in human development. One of the most pronounced aspects of the so-called Industrial Revolution was a transition to more intensive energy sources, which was essential to support wide-scale mechanization. Today, industries represent the manufacture of a number of goods vital to the functioning of modern economies, which at a primary level includes petroleum refining, steel and aluminum processing. At a secondary level, cement, paper, and chemicals industries are also vital to robust economies, requiring refined petroleum products, steel, aluminum, electricity, and transportation to function. Clearly, each and every manufactured product contains a large degree of embedded energy. This is an unavoidable consequence of modern economic activity.

1.4.2.4 The Residential and Commercial Sectors

As stated earlier, residential and commercial sectors typically are lumped together because a number of mixed-use activities occur in our homes and offices in similar ways. The most significant differences in residential and commercial energy uses amongst population demographics will involve increased heating and cooling needs based on climate, orientation to the sun, and fuel type, which itself can be based on locally available and/or utilized fuels. In addition, residential and commercial structures will be depended on for essential activities like lighting, heating and pumping water, operating appliances for refrigeration, communication, and entertainment. The systems for internal climate control and lighting and household appliances can remain in place for several years. Communication, computation, and entertainment technologies can be turned over on the order of just a few years. As populations continue to urbanize, as countries shift to more service-based activities, and as consumers continue to utilize energy intensive technologies in everyday life, the structures where we live and work likely will become larger consumers of fuels and electricity.

1.4.3 ENERGY USE BY EXAMPLE: THE UNITED STATES

As of this writing, the United States, with about 5% of the world's population, represents about 22% of the world's economy and about 22% of total global energy use. This is down from a high of about 25%. Energy consumption, tied to abundant and locally available natural energy resources, has been instrumental in providing this country's fantastic rise to affluence and preeminence in the world's economy and society. Much of this energy use is tied to a growth-based economy where consumption is an essential component.

1.4.3.1 Consumption at the End of the Last Decade

Let us take a detailed look at United States energy use by source (supply) and by sector (demand) in 2009, which is depicted below in Figure 1.3. First, note that 2009 United States consumption was 94.4 Quads. (Recall that a Quad is shorthand for one quadrillion BTUs, which is equivalent to 172 million barrels of oil or about 10 days of United States oil consumption in 2009.) Compared with average annual consumption in the first decade of the 21st century, which was as high as 101 Quads in 2007, the consumption in 2009 was relatively low, indicative of the "poor" state of the United States economy.

We see that on the supply side of Figure 1.3, oil and gas together accounted for about 58.7 Quads, or 62% of total consumption in the United States. When coal is added to oil and gas, this number climbed to 78.4 Quads, or 83% of total consumption—it is clear that the fossil

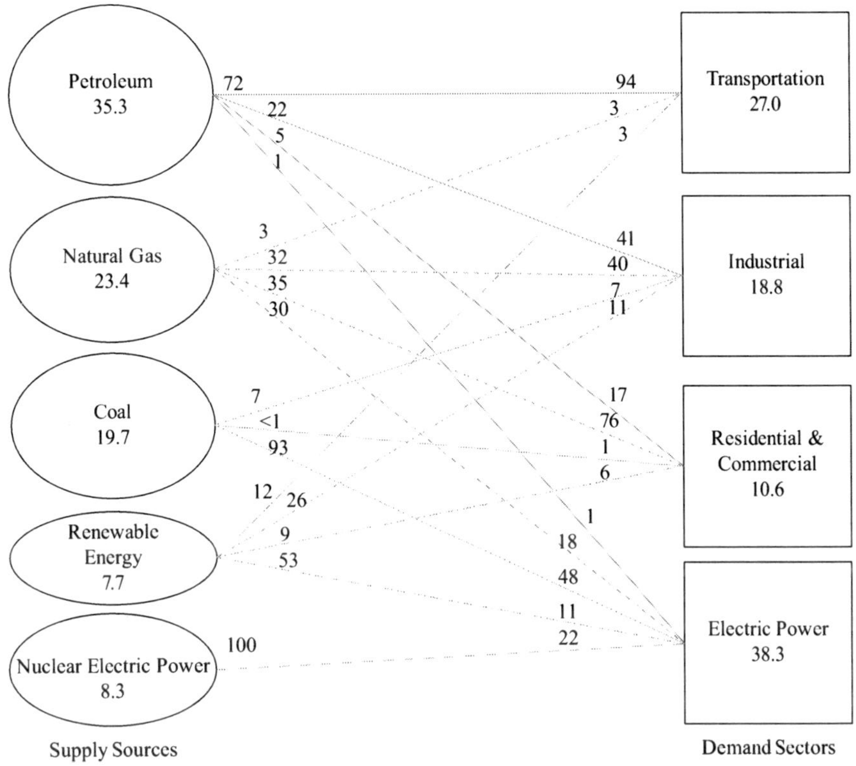

Figure 1.3. 2009 United States Energy Source/Sector Linkages.
Source: eia.gov

fuels utterly dominate the marketplace of one of the largest economies of the world. After the fossil fuels, nuclear electric power represented the next largest supply at 8.3 Quads or 9% of the total, which was closely followed by renewable energy at 7.7 Quads, or 8% of the total. On the demand side we see that electric power generation was the largest single demand sector requiring 38.3 Quads, or 41% of total primary energy. We also see that 100% of the nuclear electric power supply, 93% of the coal supply, 53% of the renewable energy supply, 30% of the gas supply, and just 1% of the oil supply went to this sector. Multiplying these percentages by Quads of energy demanded we see that just a little less than half of the electric power sector was dependent on one resource: coal (48%). This was followed by nuclear power (22%) and then gas (18%), with renewable electricity a distant fourth (11%). The next most demanding sector was transportation at 27 Quads and this story was much more one-sided. Whereas only 72% of the petroleum supply went to the transportation sector, when multiplied by total energy demanded, oil represented a whopping 94% of that sector. The industrial sector, demanding 18.8 Quads was evenly dependent on oil (41%) and gas (40%), with a smaller dependence on coal and renewable energy, presumably in the form of heat. Finally the residential and commercial sectors demanded 10.6 Quads and were heavily dependent on gas (76%) for heating.

While Figure 1.3 is quite informative, it fails to relay the waste associated with the consumption of energy. For that, we turn to the Lawrence Livermore National Laboratory (LLNL) and their rather extensive breakdown of the energy sources that are processed, consumed, and wasted in the United States economy, provided in Figure 1.4. This Daedalian maze depicts the

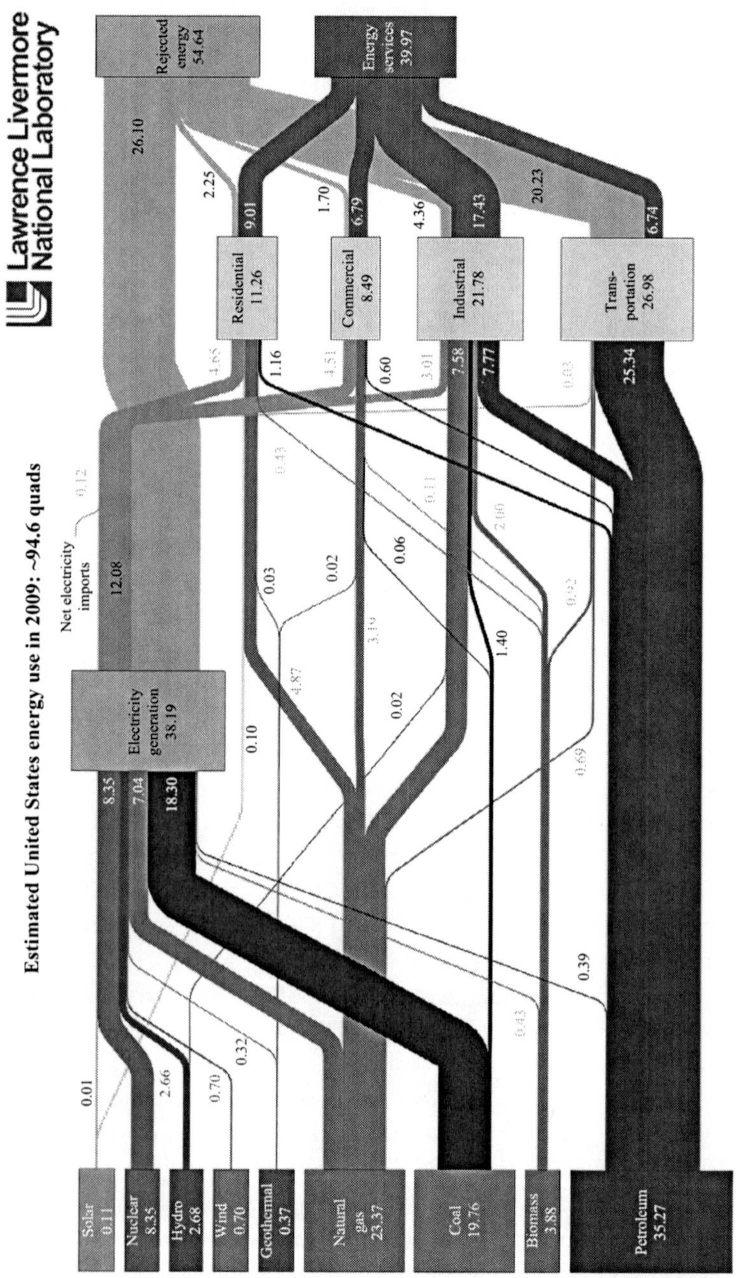

Figure 1.4. 2009 United States Energy Flow Diagram.
Source: LLNL (2010).

intricacy of energy flow to and in the United States. Again, energy flows from sources into consumption sectors, with resulting useful outputs. Waste streams of "rejected energy" total 54.6 Quads or 58% of the total source inputs. It is useful at this point to explain that waste is inevitable, given the constraints of the first and second laws of thermodynamics. The first law states that energy can neither be created nor destroyed, just converted from one form into other forms. This gives us some assurance that the primary energy can be manipulated to make available useful energy. The second law—the law of entropy—essentially says that the availability of this useful energy can only dwindle. The largest culprit, of course, is heat. It is possible to use the resources more efficiently to reduce these losses, but waste will always exist.

1.4.3.2 Renewable Energy's Role

Concerns over the long term availability of fossil and nuclear resources over the past several decades, as well as concerns over the very polluting nature of coal combustion for electricity generation, have placed greater emphasis on the increase of renewable energy capacity in the United States and in the world. (This has been of greater concern to other regions, particularly Western Europe, where local availability of resources is much scarcer than in the United States) And while there has been a concerted effort (and recent progress) to increase capacity over the past decade in particular, renewable energy still is dominated by a couple of major resources, and of those that dominate, all are utilized as stock-limited energy.

Figure 1.5 depicts the renewable energy supply for the United States from 2005 to 2009 (EIA 2010). Notice that the United States renewables supply increased from 6.4 Quads to 7.8 Quads over the five year span, where in this span, and indeed historically, the contributions of renewable sources have been impressively dominated by hydropower, and wood and wood-derived fuels. In the contiguous 48 states of the United States, large-scale, dam-based hydropower has been extensively developed and has marginal opportunity for expansion. Wood energy has the potential to grow in contribution, but most substantially in the form of liquid biofuels, which indeed have grown dramatically in recent years, but in the form of corn-based ethanol. Mandated to be used in motor fuel, ethanol has allowed biofuels to grow from 0.5 Quads

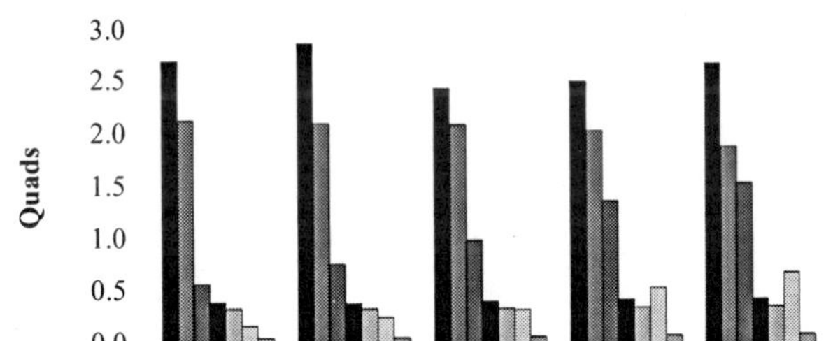

	2005	2006	2007	2008	2009
■ Hydroelectric conventional	2.70	2.87	2.45	2.51	2.68
■ Wood and derived fuels	2.14	2.11	2.10	2.04	1.89
■ Biofuels	0.58	0.77	0.99	1.37	1.55
■ Waste	0.40	0.40	0.41	0.44	0.45
■ Geothermal energy	0.34	0.34	0.35	0.36	0.37
■ Wind energy	0.18	0.26	0.34	0.55	0.70
■ Solar thermal/PV energy	0.07	0.07	0.08	0.10	0.11

Figure 1.5. United States Renewable Energy Supply from 2005 to 2009.
Source: eia.gov

to over 1.5 Quads in just 5 years. Corn ethanol's increased production likely will be short lived as ethanol from cellulosic material has the most potential for continued (theoretical) expansion.

Wind power generation also grew very rapidly over the time period shown in Figure 1.5, with a remarkable near quadrupling in just five short years. And if we look at contributions in 2010, 2011, and 2012, wind power continues to grow rapidly and now contributes over one Quad of energy to the energy supply annually (EIA 2013). This is one of the most encouraging observations for renewable sources at the time of this writing. Still, it should be noted that it will need to increase its production 20 more times to match the energy obtained from coal in the United States, in 2009. And this is without even taking into account the fact that it is a flow-limited resource, needing massive breakthroughs in storage technologies to be used in the country's economy in a way that remotely resembles current patterns, if it were to be used as a primary electricity source.

1.4.3.3 Dependence on Foreign Oil

A popular statistic often heard about the United States and its oil consumption is, "The United States, with only 5% of the world's population, consumes about 25% of the world's petroleum resource." While true (well mostly—remember that the consumption percentage is down a bit as of 2010), two important points should be made. The first point is that the United States has represented about 25% of the world's economy over several years, so that consumption statistic in itself should not be too surprising. (Although, this purported economic activity has been tied to voracious consumption that is now buried in foreign debt). The second point is that the United States is still the third largest producer of petroleum (including petroleum-derived liquids) in the world. This second point is actually pretty important, because quite often it is perceived that the United States is almost exclusively dependent on foreign oil. It is not uncommon to hear people say that foreign oil represents 80% to 90% of total United States consumption. But if we look at the historical trends in Figure 1.6 we see that there is still some time longer before that large of a percentage is realized, although, how much longer is subject to debate. Indeed, as we write this book, liquid production (oil) from the shales is growing, especially in the United States.

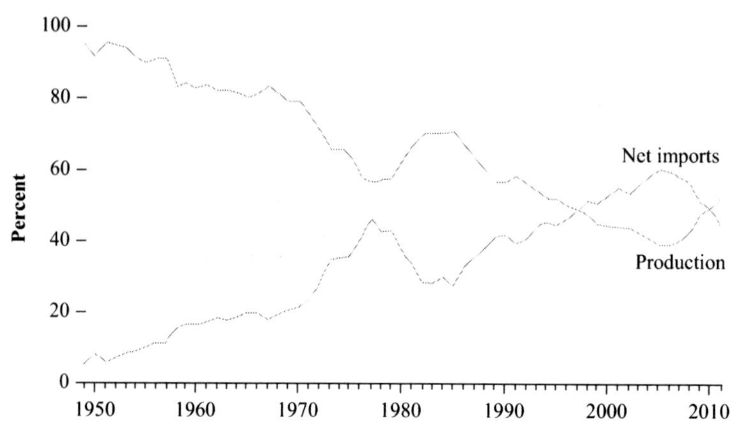

Figure 1.6. Total production and net imports as a share of consumption from 1949 to 2011. Net imports represented 45% of total United States consumption in 2011.
Source: eia.gov

It appears feasible that the United States could become independent of imported oil by about 2030. Of course, this oil production from the shales will peak and decline too, but the seemingly inevitable growth of United States import-dependence is not as inevitable as one may think.

Figure 1.6 should nonetheless give pause to the unsustainable pathway the United States has been following with respect to its oil dependence. In 2005, this trend reached a local maximum of 60.3% but declined to just about 45% in 2011. Whether this decline is real and can be sustained is unclear. Shale plays and a supposedly revitalized economy will determine how this trend progresses. Either way, the dependence on foreign oil is an issue that will continue to plague the United States and practically every other industrialized nation, as most future oil will come from regions that fall outside of their jurisdiction. Indeed British Petroleum's 2007 *Statistical Review of World Energy* estimates that 76% of proven oil reserves exist in countries represented by the Organization of Petroleum Exporting Countries (OPEC), which includes nations in Latin America, in the Persian Gulf, and in Africa. This will be an issue that will not go away for the foreseeable future.

1.5 ENVIRONMENTAL IMPACTS OF ENERGY USE

Rational evaluations of energy use in societal structures such as those found in Smil (2003), Khodeli (2009), and UNDP (2005) make it obvious that humanity has benefitted greatly from the use of fossil and nuclear fuels as primary energy sources. However, it is equally obvious that the scales with which we use them have resulted in a number of negative impacts on Earth's atmosphere, waterways, and ecosystems. These impacts can often be quantified in terms of change to existing natural systems, as well as in the monetary costs associated with impact on society, including those related to health.

As per Boyle (2004), there are three general classifications of the impacts of energy use: (1) classification by source; (2) classification by pollutant; and (3) classification by scale. While this is a rather robust way to look at impacts, we choose here simply to describe pollution and group sources by what kinds of pollutants or "harms" they create and by the scale to which they affect the environment.

1.5.1 CLASSIFICATION BY POLLUTANT OR HARM

We process raw energy for utilization in economic sectors mainly by combustion, by nuclear fission, and by the capture of kinetic energy from renewable flows for electricity. The impacts of the processes, in the forms of pollutants or harms released to the surrounding environment, can subsequently be classified in the following ways: air pollution; water pollution; radioactivity; land use and ecosystem change; noise and aesthetics; and climate change. Each is described briefly below.

1.5.1.1 Air Pollution

Air pollution is defined as the emission of gases and particulate matter into the environment. These emissions can cause discomfort or harm to humans, and can cause damage to the natural environment, whose effects can be felt globally once entering the atmosphere. Probably the most pervasive source of air pollution is from the production of airborne substances during the combustion of fossil and biomass fuels. Some notable pollutants that are released when fuel is burned include: (1) carbon monoxide, which can be highly toxic to humans and animals even at low concentrations; (2) sulfur dioxide, which causes acid rain and respiratory illnesses; (3) nitrous oxides and volatile organic compounds, which can irritate and damage organs directly and also can react in the atmosphere to produce the secondary pollutant ozone, which itself is harmful to lung function; (4) particulate matter, which causes hazy conditions in cities, and can contribute to asthma and lung cancer at very small sizes; and (5) heavy metals such as lead and mercury, which can affect neurological and development functions. Finally, carbon dioxide (CO_2) has been targeted as an amplifier of atmospheric warming, owing to its ability to absorb infrared radiation. In addition to fossil and biomass energy, hydropower is also a source of this emission.

1.5.1.2 Water Pollution

Water pollution is defined as the emission of any matter that can enter into waterways directly or indirectly through the hydrologic cycle, and can also include thermal emissions. Indeed, as discussed in Masters and Ela (2008), any time water is withdrawn, used, and returned to a source, it is likely polluted. There are numerous sources of water pollution, but those derived from energy use include (1) the discharge of liquid fuels (purposeful or otherwise) during extraction, refining and end use; (2) the discharge of higher temperature water from fossil and nuclear power plants; (3) the deposit of mercury from air (through the hydrologic cycle) after being emitted by coal combustion; and (4) the discharge of oxygen-deprived and/or temperature differentiated (hot or cold) water from hydropower plants.

1.5.1.3 Radioactivity

Radiation is the chief environmental concern related to fissile fuels and their products, used in nuclear power generation. In particular, it is ionizing radiation that presents the special hazards of concern in nuclear power: that is radiation with sufficient energy to dislodge electrons from atoms and molecules. Radiation hazards are measured in rads, rems, and several other units. A rad refers to the amount of energy imparted to living matter by radioactivity, while a rem refers to the amount of ionizing radiation. Radiation is produced as radioactive isotopes either undergo fission or natural decay. Every radioactive isotope has a measurable half-life, the amount of time it takes for half of the isotope to decay to its daughter product(s). The length of the half-lives may be as little as a tiny fraction of a second, or many millions of years. The half-life of potassium-40, which naturally occurs all around us, is over one and a quarter billion years, while the half-life of uranium-238 (the most abundant natural form) is almost four and a half billion years. Although the long half-lives of some of the fission products are often cited as a reason for fear, the reality is that many chemical toxins do not lose toxicity over time at all. When a half-life is very long, it indicates that its hazards cannot be considered to diminish in a realistic time frame, much like common chemical toxins.

1.5.1.4 Land Use and Ecosystem Change

Large-scale operations of all kinds require extensive resources, particularly land. However, when land is transformed for human use it often disrupts ecosystems, whether through flooding and desertification, through clearing for crops, or by contamination. Urbanization and urban sprawl are two of the most tangible contributors to land use changes that can negatively impact the environment. Especially without mass transit, roadways necessary to transport people and goods disrupt ecosystems. Impact on ecosystems also follows from all of the other kinds of impacts described earlier. Air pollution affects the health and well-being of all living things. Air pollution has led to acidification, which alters the chemistry of waters and soils. Water pollution impacts all kinds of aquatic life. Even the construction of large wind farms in pristine environments has generated concern. The intricacy of the balance of ecosystems makes this kind of impact particularly challenging to mitigate. Certain species depend on very specific other species for their sustenance—or even reproduction, as some plants are pollinated by specific species of insects. When one piece of the ecosystem is removed, it affects others and an imbalance can result.

1.5.1.5 Noise and Aesthetics

Aesthetic concerns can be conflated with environmental impacts. Noise has particular standing in this regard, as evidenced by the term noise pollution. Noise may affect some distribution of species, as some animals may avoid certain kinds and levels of noise. Noise can be quantified, unlike many of the other aesthetic concerns. At approximately the level of busy city traffic, 85 decibels, a person subjected to more than eight hours of continuous exposure is likely to have some hearing loss, with permissible exposure times reduced by approximately one half for every additional three decibels. Some 10 million Americans are estimated to have noise-related hearing loss. Other health issues from noise include sleeplessness, high blood pressure, and lost productivity (EPA 2012; CDC 2012). There seems to be, however, little lasting impact related to other aesthetic concerns.

1.5.1.6 Climate Change

Climate change is defined as any significant change in measures of climate (such as temperature, precipitation, or wind) lasting for an extended period (decades or longer). It may result from natural processes involving the sun–earth linkage as well as from human activities that change the earth's surface or composition of the atmosphere. Both natural and human climate triggers result in climate changes responding to these triggers, often in the form of warming or cooling. Humans have more than likely impacted the climate system for many thousands of years, dating back to the time when intensive agricultural practices were adopted to better secure food supply (Ruddiman 2010). Of recent particular interest is the emission of CO_2, methane, and nitrous oxides from combustion and from industrial agriculture. As strong infrared absorbers, many argue it is highly likely that the emission of these and other "greenhouse gases" has amplified and/or caused a period of global warming in recent times. Assessing the potential harm that these emissions could cause on the environment and humanity often involves the modeling of highly complex processes, which can be quite difficult. The very nature of models is that they cannot be perfectly accurate, especially when applied to local conditions. The only way we will be able to determine with absolute accuracy the actual climate change phenomenon is to wait until it has played out and generated sufficient data to prove (or disprove) key parts of the models. No matter how skeptical anyone may be of the evidence for climate change, it behooves humanity to take measures to mitigate its severity. Most of those measures will have salutary effects on energy conservation and environmental preservation anyway. Indeed, more sustainable energy systems will almost certainly have smaller "carbon footprints." To some extent, climate change is a bellwether for other environmental problems described above.

1.5.2 CLASSIFICATION BY SCALE

Holdren and Smith (2000) provide a comprehensive assessment of environmental impact by scale, but briefly, impacts may be localized to households, workers, communities, regions, or they may be global in scale. The production of smoke and other pollutants in domestic biomass burning is considered a household-scale impact. The hazards of coal mining almost exclusively impact the workers, while contamination of a shallow aquifer primarily impacts the community. Air and most surface water contamination have been seen to have regional impacts. Climate change is clearly an impact at the global scale.

1.6 DEFINING SUSTAINABILITY AND SUSTAINABLE ENERGY

In this section we will discuss the concept of sustainability and what it means to energy use in this century. It is important to note that sustainability and sustainable development often are used interchangeably. Development, in this context, is any activity that is undertaken for the purpose of economic or social benefit.

1.6.1 SUSTAINABILITY

Building on a definition presented by Davidson et al. (2007), sustainability is designed to help actors in society determine the likely, plausible, and potential outcomes and changes to economic (and thus resource), environmental, and social systems that result from a particular development decision (e.g., the making and selling of goods, or the providing of electricity to homes). This draws on the concepts of sustainable development laid out in the Brundtland report, *Our Common Future* (1987), which ultimately aims to achieve and maintain an earth-human system that provides adequately for present and future generations. Addressing these goals is an optimization exercise that seeks to maximize desired economic or social outputs while minimizing economically, socially, and ecologically costly inputs. This makes for a broad framework on which to build, but one with a wide range of approaches and possible understandings. It is useful then to define its characteristics in terms of two divergent perspectives, known as "weak" and "strong" sustainability, and then to proceed with what sustainability means in practice for energy.

1.6.1.1 Weak Sustainability

According to Ayres (2007), weak sustainability stipulates that as long as future generations acquire stocks of natural, man-made, or human capital (e.g., a tree, a watch, or a person) equal to or greater than those stocks available to the present generation, sustainability can be satisfied.

This is true even if this activity is done by completely exhausting the stocks of natural capital, as long as those stocks are able to be *substituted* by at least one of the other two. In essence, weak sustainability assumes that man-made capital (goods, services, money) is equivalent to natural capital and, thus, can be valued equally and in distinct monetary terms. It is argued this can be done because technological innovation and breakthroughs allow it to be done. The history of illumination is a classic example that Ayres uses in support of this sustainability type, as humans have substituted animal fat torches for candles, then oil lamps, then kerosene lamps, then gas lighting, then incandescent lighting, then fluorescent and LED lighting. The substitutions seem endless.

It should seem intuitive that weak sustainability is a flawed concept. As Ayres points out, it neglects notions of "mineralogical barriers" to extraction. Technology is often dependent on physical resources to make it function. If those resources are exhausted, the technology can no longer function. Additionally, there is this problem of determining an acceptable value, monetary or otherwise, to a natural system. A good example is a forest, which serves as a source for materials but also as a source for biodiversity, habitat, and aesthetic appreciation.

1.6.1.2 Strong Sustainability

Again, following Ayres, strong sustainability, which is also concerned with keeping capital stocks constant over generations, adds to weak sustainability the requirement that stocks of natural capital in particular should not be diminished. In this definition, then, these stocks are to be reserved as "ecological assets" that are not replaceable. Simply put, strong sustainability acknowledges that we live on a finite planet, where "stocks of natural resources, as well as many ecological functions, are irreplaceable," and therefore not substitutable. No degree of man-made capital can be used to re-create the hydrologic cycle or the ozone layer, for example. According to Ayres, "under strong sustainability criteria, minimum amounts of a number of different types of capital (economic, ecological, and social) should be independently maintained, in real physical/biological terms." The classic example used to support this sustainability type is the current rapid depletion of the fossil fuels coal, oil, and gas. The fossil fuels represent millions of years of accumulated solar energy and our current consumption rates will deplete this stored solar energy in hundreds of years—a rate that is much faster than is practically replenishable.

It should seem obvious that strong sustainability also is a flawed concept, primarily because strong sustainability conflicts with the social and economic goals of all sovereign nations. To reserve natural capital as ecological assets, and thus untouchable, is not consistent with economic endeavor. We live on a world dependent on the utilization of natural capital for societal benefit. A compromise is needed.

1.6.1.3 Sustainability in Practice

The problem with sustainability comes in its application in the real world. Decision makers often have their own local self-interests at stake and may make decisions that are helpful at the local level, but that are detrimental (i.e., not sustainable) at the global level. Moreover, those decision makers may have resources available to them locally that are not available globally, making the sustainability of that decision relative to location. On top of that, a decision may be deemed sustainable over a 30-year time frame but unsustainable over a 100-year time frame, making time an important factor in decisions where the stated underlying goal is to sustain a system for future generations. On top of that, pollution affects sustainability because the impacts of a local decision can be exported to other locales, either by importing goods and the inherent resources they required to manufacture and transport or by transferring emissions via the wind or the water, which can harm ecosystems and humanity. Ultimately, sustainability can only be relative, i.e., choices can be made to minimize unsustainability, not to establish perfect sustainability.

Despite these shortcomings, what sustainability as a concept does is raise the issue that decisions at every level ultimately affect the environment and society in some noticeable way, more so as the influences of any form of consumption overwhelm the system—especially as limits of physically available resources are approached. Thus a good compromise is one that acknowledges our effect on economic, ecological, and social systems and that strives to preserve or maintain, again according to Ayres, "a minimum amount of certain environmental assets, based on the idea that these assets are partly complementary to economic assets and partly substitutable by the latter." It is this compromise that must define sustainable energy pathways.

1.6.2 SUSTAINABLE ENERGY

The current global energy system is structured around the use of high quality, highly transportable, nonrenewable fossil and nuclear fuels. The level of consumption of these resources has reached the point that its use represents some conspicuous challenges to the long-term viability of essential natural and human services that exist today. With the reality that fossil and nuclear fuels are finite, are consumed at extremely rapid rates, and emit large amounts of pollutants when consumed, humanity must learn to choose other options that might help maintain high standards of living in a more sustainable world. Defining sustainable energy, then, becomes very important, and requires a framework that includes supply side and demand side expectations for energy use.

1.6.2.1 Sustainable Energy Supply

Any sources that are to be responsible for the raw, energetic supply of the world's economy must be robust enough to allow for new development activities and to sustain existing developments. This means that they must be able to adequately supply the five economic demand sectors described in Section 1.4. Now, if these sources also are to adhere to sustainability, then, according to Boyle (2004), they must meet the following criteria:

- They are not depleted by continued exploitation.
- They do not cause the harmful emission of pollutants or other negative impacts on the environment.
- When used, they do not cause human health hazards.
- The harnessing, transporting, and trading of them does not cause and/or perpetuate social injustices.

If we consider the collective fossil and nuclear fuels, it is evident that all of these energy sources fail to meet all of the above criteria. However, it can be argued that technology has improved their relative sustainability in many regards, just not sufficiently to strictly (perhaps even remotely) meet Boyle's prescription. The so-called renewable energy sources do a better job at meeting more of these criteria, but by no means do they adequately meet them all either. Moreover, once we go into the finer details of stock versus flow-limited energy, we see that the sustainability of certain renewable energy supplies can be questioned, as they can be depleted under certain circumstances (e.g., biofuels). Additionally, since all energy sources require that a technology interface with them in order to create the appropriate energy service, it is clear that no supply can be truly considered sustainable, because those technologies likely depend on finite mineral resources for their construction.

A more forgiving definition of a sustainable energy supply is one that (1) maximizes its availability to be exploited by present and future generations; (2) does not substantially compromise long term environmental assets, either in the form of excessive harmful pollutant emissions or by diminishing other resources that are used in their production and delivery; (3) does not substantially compromise human health; and, (4) one that is widely accessible to the entirety of the world's population. This definition opens the door for renewable and responsibly used nonrenewable resources, but only if used in a more equitable mix than currently exists and one that can phase out the least sustainable sources over time. This represents a grand challenge for humanity in this century moving forward.

1.6.2.2 Sustainable Energy Demand

Energy has been and will continue to be central to the functioning and improvement of human societies. Shaping its procurement and perhaps more importantly its consumption are the key driving forces for devising energy strategies to achieve sustainable energy in this century. Thus, sustainable energy demand is defined simply as energy demand that must maximize both the efficiency and sufficiency of use. As such, energy efficiency and energy sufficiency connote two distinct solutions: the first is technology-based and the second is behavior-based.

On the technology front, a recent study by Cullen et al. (2011) suggests that with existing technologies roughly 70% of global energy use "could be saved by practically achievable design changes to passive systems." These changes include everything from the glazing of building windows, to the insulation of building wall cavities, to the limiting of the weight of automobiles for reducing transportation energy demands. Although this does not speak to the actual likelihood of implementation of these practices, we see that, if implemented, the sustainability of the demand side would be impacted dramatically, given the reduction in energy usage through better efficient practices with existing technologies.

On the behavior front, there is this relatively new sense for determining what constitutes "enough energy" to maintain high standards of living for the entire world population. A project initiated by the Swiss Federal Institute of Technology (SFIT) suggests that the world could theoretically become a "2000 watt society" by 2050, in which each person on the planet could consume energy at a rate of 2000 watts without compromising a high quality of life (SFIT 2008). 2000 watts is roughly the world per capita average today. This would require a massive redistribution of usage, however, which likewise would require massive incentives for heavily consuming nations to reduce consumption. Again, this does not speak to the actual likelihood of implementation, but it does suggest that all of the energy we need to maintain high standards of living is available today.

1.6.2.3 Sustainable Energy in Practice

Sustainable energy should be viewed as a relative concept, surely not absolute. That is, certain energy sources in certain contexts and over certain time frames will meet sustainability criteria better than do others, given the needs of a particular development activity. No resources or technologies will meet all of the sustainability criteria perfectly or indefinitely. And determining the relative sustainability of one energy system over another is highly dependent on the location, the technologies to be used, and the values and interests of the primary stakeholders involved (Boyle et al. 2003). This could mean nonrenewable or renewable resources or various combinations of them. And while we need to be pro-active about the reduction of the use of fuels that create the most destruction to environmental assets, in practice, it is our ability to deplete the reserves of some resource that limits the sustainability of its use in the future. And, ultimately, it is both the supply and the demand that totally defines sustainable energy.

1.7 SOURCES OF ENERGY AND ENVIRONMENTAL INFORMATION

There are a number of sources that produce information on consumption, production, and sources of primary energy and on environmental data related to energy use. For example, investor-owned international energy companies provide some of the most comprehensive energy information publicly available. But often, as Shaffer (2011) notes, we turn to governments and how they publish energy data with analysis specific to the conditions and policies of their state. As major actors in the purchasing, processing, and overall flow of energy resources, these entities inevitably focus on trends that are important to their respective constituencies, even if the information agency producing the analysis is protected from direct governmental approval or influence. Finally, as governments have taken up the responsibility to protect the environment and society from emissions of various pollutants that result from energy use, a number of additional entities, both governmental and nongovernmental, have been created to report on additional data related to energy use impacts.

1.7.1 UNITED STATES ENERGY INFORMATION ADMINISTRATION

The EIA originated from the Federal Energy Administration Act of 1974 in response to the energy crises of that time. In 1977, under the Department of Energy (DOE) Organization Act, the EIA was established "as the single Federal Government authority for energy information" (eia.gov). The EIA is independent from the DOE and does not require approval from the United States government for data collection and for the issuing of reports. In addition to detailed annual energy statistics available publicly, the EIA issues three popular "outlook" reports: the *Short-term Energy Outlook*; the *Annual Energy Outlook*; and the *International Energy Outlook*.

1.7.2 INTERNATIONAL ENERGY AGENCY

Much like the EIA, the International Energy Agency, or IEA, was established during the 1973–1974 oil crisis "to help countries co-ordinate a collective response to major disruptions in oil supply through the release of emergency oil stocks to the markets" (iea.org). With 28 member countries, the IEA is more global in reach and serves to provide research and analysis and recommendations for its members to procure affordable and reliable energy. The IEA is an autonomous agency with members from North America, Europe, and Asia, which also generates important reports on energy, perhaps the most popular being the *World Energy Outlook*.

1.7.3 WORLD ENERGY COUNCIL

The World Energy Council, or WEC, is a nongovernmental, member-based organization with an interest in understanding emerging energy issues. WEC has member committees in 93 countries and their stated mission is "to promote the sustainable supply and use of energy for the greatest benefit of all people" (worldenergy.org). WEC organizes a world congress every three years and issues a number of reports every year on energy data and analysis. One of their more notable publications is the annual *Survey of Energy Resources*.

1.7.4 WORLD RESOURCES INSTITUTE

The World Resources Institute, or WRI, is an influential environmental think tank based in Washington, D.C. WRI is an independent, nongovernmental organization originally established as "a center for policy research and analysis addressed to global resource and environmental issues" (wri.org). WRI is best known for its World Resources report, and actively publishes on climate and energy. WRI's *EarthTrends* environmental database, although no longer actively adding data, remains an extremely useful source of energy data.

1.7.5 INTERGOVERNMENTAL PANEL ON CLIMATE CHANGE

The Intergovernmental Panel on Climate Change, or IPCC, was established by the United Nations Environment Programme and the World Meteorological Association. It is the framework by which scientists and other participants digest the current literature on climate change and present an overview and assessment of the most relevant findings. The work of the IPCC is "shared among three Working Groups, a Task Force, and a Task Group" (ipcc.ch). Working Group III (WG III) is the one most relevant to energy, as it assesses the current options available for mitigating climate change, particularly through limiting, preventing, or removing emissions from the atmosphere. Since emissions are mostly a product of energy resource consumption, it is useful to be aware of IPCC WG III's ongoing work.

1.7.6 INDUSTRY REPORTS

British Petroleum, or BP, Royal Dutch Shell, and ExxonMobil are examples of investor-owned energy companies that are driven by maximizing shareholder returns. As such, they quickly develop and produce oil, gas, and other energy resources to which they have access and sell their production output in competitive international markets. Assessing the current and future availability of resources is important to their viability, thus collecting relevant data to inform investment decisions is compulsory. Three widely available and oft used reports issued by these companies include BP's *Statistical Review of World Energy*, Shell's *Energy Scenarios to 2050*, and ExxonMobil's *Outlook for Energy*. These reports are cited heavily in most governmental and nongovernmental reports.

REFERENCES

Ayres, R. U. 2007. On the practical limits to substitution. *Ecological Economics 61* (1): 115–128. doi: http://dx.doi.org/10.1016/j.ecolecon.2006.02.011

Beiser, A. 2009. *Schaum's Outline of Applied Physics*. New York: McGraw-Hill.

Boyle, G., B. Everett, and J. Ramage. 2003. *Energy Systems and Sustainability*. New York: Oxford University Press.

Boyle, G., ed. 2004. *Renewable Energy*. New York: Oxford University Press.

Center for Disease Control (US CDC). 2012. *How to Look at Noise?* http://www.cdc.gov/niosh/topics/noise/chart-lookatnoise.html

Cullen, J. M., J. M. Allwood, and E. H. Borgstein. 2011. Reducing energy demand: what are the practical limits? *Environmental Science and Technology* ***45*** (4): 1711–1718. doi: http://dx.doi.org/10.1021/es102641n

Davidson, C. I., Matthews, H. S., Hendrickson, C. T., Bridges, M. W., Allenby, B. R., Crittenden, J. C., and S. Austin. 2007. Viewpoint: Adding sustainability to the engineer's toolbox: A challenge for engineering educators. *Environmental Science and Technology* ***41*** (14): 4847–4849. doi: http://dx.doi.org/10.1021/es072578f

EIA (US Energy Information Administration). 2010. *Trends in Renewable Energy Consumption and Electricity*. http://www.eia.gov/renewable/annual/trends/

EIA (US Energy Information Administration). 2011. *Annual Energy Review*. http://www.eia.gov/totalenergy/data/annual/index.cfm

EIA (US Energy Information Administration). 2013. *Monthly Energy Review*. http://www.eia.gov/totalenergy/data/monthly/#renewable

EPA (US Environmental Protection Agency). 2012. *What is Noise Pollution?* http://www.epa.gov/air/noise.html

Hern, W. M. 1999. How many times has the human population doubled? Comparisons with cancer. *Population and Environment* ***21*** (1): 59–80. doi: http://dx.doi.org/10.1023/A:1022153110536

Holdren, J. P., and K. R. Smith. 2000. Energy, the environment and health in *World Energy Assessment: Energy and the Challenge of Sustainability*. New York: UNDP.

IEA (International Energy Agency). 2011. http://www.iea.org/about/index.asp

IPCC (Intergovernmental Panel on Climate Change). 2007. *Climate Change 2007: Working Group III Report "Mitigation of Climate Change."* http://www.ipcc.ch/publications_and_data/ar4/wg3/en/contents.html

Khodeli, I., ed. 2009. *From Curse to Blessing? Using Natural Resources to Fuel Sustainable Development*. Hoboken: Wiley-Blackwell.

Masters, G. M., and W. Ela. 2008. *Introduction to Environmental Engineering and Science*. Upper Saddle River: Prentice-Hall.

Meadows, D. H. 2008. *Thinking in Systems: A Primer*. White River Junction: Chelsea Green Publishing.

Reist, P. C. 1993. *Aerosol Science and Technology*. New York: McGraw-Hill.

Ruddiman, W. F. 2010. *Plows, Plagues, and Petroleum: How Humans Took Control of Climate*. Princeton: Princeton University Press.

Shaffer, B. 2011. *Energy Politics*. Philadelphia: University of Pennsylvania Press.

Smil, V. 2003. *Energy at the Crossroads, Global Perspectives and Uncertainties*. Cambridge, MA: MIT Press.

Smil, V. 2006. 21st century energy: Some sobering thoughts. *OECD Observer*, 258/259: 22–23.

Smil, V. 2008. *Energy: A Beginner's Guide*. Oxford: Oneworld Publications.

Sorensen, B., P. Breeze, T. Storvick, S. Yang, A. da Rosa, H. K. Gupta, and S. Kalogirou. 2009. *Renewable Energy Focus Handbook*. Waltham: Academic Press.

SPE (Society of Petroleum Engineers). 2010. "Petroleum Resources Management System Guide for Non-Technical Users." http://www.spe.org/industry/docs/PRMS_guide_non_tech.pdf

SFIT (Swiss Federal Institute of Technology). 2008. *The 2000-Watt Society.* http://www.novatlantis.ch/en/2000watt.html

UNDP (United Nations Development Programme). 2005. *Energy Services for the Millennium Development Goals.* New York: UNDP.

WEC (World Energy Council). 2011. http://www.worldenergy.org/about_wec/

World Commission on Energy and Environment. 1987. *Our Common Future*. New York: Oxford University Press.

WRI (World Resources Institute). 2010. *EarthTrends.* http://earthtrends.wri.org/

CHAPTER 2

"NONRENEWABLE" ENERGY RESOURCES

This chapter discusses fossil and nuclear energy resources, the resources that comprise the nonrenewable energy supply.

2.1 FOSSIL FUELS

Fundamentally, fossil fuels are biomass that has accumulated, been buried under rock sediments, and transformed by heat, pressure, and time. Prevailing geologic theory characterizes fossil fuels as the carbon-laden remains of long dead organisms (hence the term "fossil"). The processes of death, decay, natural burial, and geologic metamorphoses are at work today as they have been for many millions of years. This constitutes a form of natural refining in which water is driven out, while the fuel is compacted and intensified. The resulting product has greater energy and power densities than the original biomass, although not quite as much total energy content.

2.1.1 OIL AND GAS

Oil and gas collectively account for more than half of the world's energy supply at the beginning of the 21st century (IEA 2010). They represent a range of products formed by the same natural processes and they typically occur together in varying combinations. Contrary to popular imagery of dinosaurs, oil and gas are generally believed to be the remains of a host of small, aquatic creatures (Ebenhack 1995). The requirements for an oil or gas reservoir would be a geologic basin in which: (1) rock sediments were deposited in an environment rich in aquatic life; (2) water was moving slowly enough not to disturb the dead or dying organic matter on the bottom; and (3) anaerobic (low oxygen) conditions exist so that the organic matter could decay slowly, without oxidizing. These conditions allow for a great deal of organic matter to accumulate and not be consumed by other creatures.

Then to become a conventional petroleum reservoir, this organic rich sediment must be buried under other rocks supplying heat and pressure for millions of years. In this process, the weight of rock overlying the original organic rich sediments compresses those sediments, squeezing water and organic fluids (oil and gas) out. Oil and gas are lighter than the surrounding water and tend to rise toward the surface. They will continue to rise until they reach an impermeable barrier. This barrier is referred to as a trap, which holds the oil and gas in place in the pores at the top of the permeable rock just below the trap. The "shale plays" emerging in the early 21st century are tapping into oil and gas that has not yet been squeezed out of the relatively impermeable source rock, into permeable and porous conventional reservoir rocks (EIA 2012).

In order to understand the issues of oil and gas production and depletion, it is essential to understand that the oil and gas do not exist in underground caverns, but rather are filling the pores or void spaces in rock. It helps to recognize what a complex process petroleum production is, as oil has to flow through long, narrow flow channels, competing with water and gas for space. It is perhaps a little easier to picture that the driving force for fluid flow diminishes (very slightly) with each barrel of oil produced. Thus, oil production declines throughout the life of the field, with each barrel of oil being slightly harder to get out than the previous ones. All of this is important in understanding why petroleum reserves are a fraction of the total amount of oil in place, even in well developed, highly productive fields.

2.1.1.1 Characterizing the Occurrence: Reservoirs

Petroleum reservoirs are vast geologic formations of porous, permeable rock—not subterranean lakes of oil. The pores in the rock contain oil (along with water) and the permeability of that rock is a function of the interconnection of pores, which permits fluids to flow through it. To better envision petroleum reservoirs, think of a beach. (In fact some reservoirs might actually be "paleo" beaches—"paleo" being a geological term for very old.) If you walk along a beach, the lapping waves demonstrate the porosity of the sand. As the tide comes in, waves wash over dry sand, turning it darker, as water fills the pores between the sand grains. As the waves recede, some of the water in the sand drains out, because the loose sand of a beach is extremely permeable.

Now imagine that your beach has been buried under thousands of feet of rock, compacting it into consolidated sandstone. Even though many of the grains have been cemented together at their edges by chemical processes over the millions of years of burial, there are still pore spaces between most of the sand grains. Oil and natural gas formed in underlying rocks will tend to rise because the oil is lighter than the water that is buried with the rocks. Oil and gas rise easily into porous, permeable rocks like sandstones (and limestones and dolomites). They push water out, occupying some of the pore spaces. (Oil and gas can never displace all of the water that was in the sand that was buried under the ocean, just as you can never get all the drops of water off a drinking glass without letting it dry in the air.)

Since oil and gas are lighter than water, they continue rising over geologic time until they reach some impermeable barrier, called a trap. The trap is often an overlying layer of tight shale, but it can be the flanks of a salt dome, a sealing fault, or any other kind of impermeable geologic feature. The hydrocarbons (a collective name for oil and gas) will remain in the pore spaces of the reservoir under the impermeable trap (see Figure 2.1) until some channel connects it to lower pressured overlying rocks—or to the surface. That's the role of the wellbore.

Once a well is drilled into a petroleum reservoir, it provides a conduit for the oil to flow to the surface. So, oil flows from the surrounding rock into the wellbore. The flow of oil through the rocks is driven by the expansion of oil, which generally contains some gas in solution, much

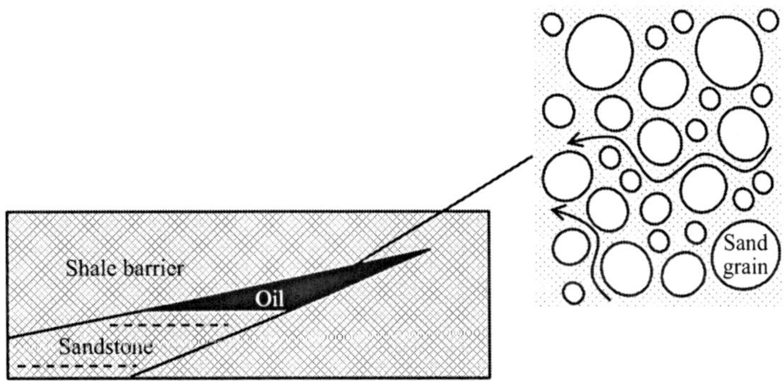

Figure 2.1. Oil deposit with a blown up depiction of oil flowing through small pore spaces. Shale formation after depiction from the Wyoming State Geologic Survey.

like CO_2 bubbles are contained in soda, beer or champagne. If you release the pressure suddenly on any of these beverages, you can see a dramatic demonstration of how fluids flow in response to decompression. The less gas there is in solution, the more challenging it is to produce oil. In the case of a natural gas reservoir, gas decompresses so effectively that the recovery factor is always much higher than for oil reservoirs.

Petroleum must flow through hundreds of feet of tortuous (crooked) pore channels to reach the wellbore. With each barrel of oil extracted, the reservoir pressure declines, thereby reducing the driving force for subsequent production. The oil and gas have been compressed by the confining pressure of the formation, so they expand as they decompress, helping to push more oil and gas to the producing wells. Because of the volumes of petroleum reservoirs, the limited permeability of the porous rock, and the compressibility of the oil and water at reservoir conditions, real fields are often able to operate at a relatively stable rate for several years before a production decline is seen. Nevertheless, an ongoing, gradual decline in the maximum production capacity of any given field is inevitable. When many fields are aggregated together, the overall production decline is more gradual, as the declines of the different fields are never perfectly synchronized, so some begin to decline while others may still be growing.

2.1.1.2 Distribution

Currently producing petroleum reserves tend to be highly concentrated in the Middle East (as seen in Figures 2.2 and 2.3). This gives many people an erroneous notion of the uniqueness of oil and gas occurrences, perhaps because the current relative abundances of these reserves are skewed to a singular region. In the early days of the petroleum business, it seemed inconceivable to many people that such hydrocarbon riches could be found anywhere in the world the way they were in Pennsylvania. One head of the United States Geological Survey notoriously offered to "drink every drop of oil they can produce from Texas" (Yergin 1991). Similar claims were made of Saudi Arabia and other places that ultimately proved highly productive. The whole history of oil and gas exploration is replete with incredulity and surprising new discoveries. Indeed, the "rediscovery" of oil and gas in the North American fields has been changing reserves estimates on what seems to be a daily basis (though totals after the proved reserves additions do not yet rival those of the Middle East).

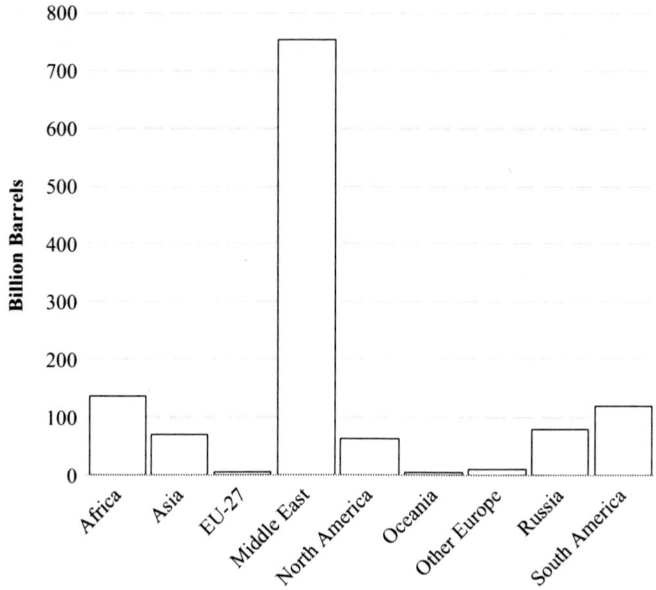

Figure 2.2. Proved recoverable oil and natural gas liquids reserves by region, end 2008.
Source: WEC (2010).

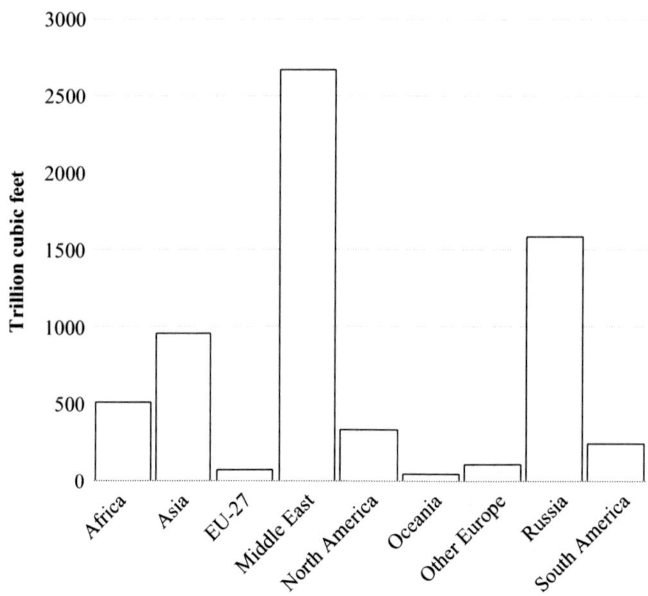

Figure 2.3. Proved recoverable natural gas reserves by region, end 2008.
Source: WEC (2010).

Those who don't understand the speculative nature of petroleum exploration are very likely to write off regions simply because no one can know what has yet to be found. This failing is not limited to the uninitiated, though. Coauthor, Ebenhack, left a technology management position in international exploration with a major oil company in 1987 to assist people in developing countries in identifying and producing their own energy resources for their own use. In the very first year of work on that project, a nongovernmental organization (NGO) in Uganda contacted Ebenhack about the need for substantial energy production to support the institution's needs and asked if there was any possibility of developing small-scale oil or gas in Uganda to that end. Ebenhack would have been delighted to assist with this project, but reviewing available exploratory data, he regretfully concluded that he could find no adequate data to support philanthropic oil or gas development efforts in Uganda. Only minimal evidence of some geologic potential in western Uganda existed. While the lack of solid evidence need not deter a major oil company with assets sufficient to support the economic losses of exploratory failures, a multimillion dollar failure could be devastating to a small nonprofit. At the writing of this book, major oil and gas discoveries have been made in western Uganda (E&P 2009). The sparse data that were available in 1988 simply did not show the potential that was there—until an oil company invested more capital in exploration and actually drilled the first productive wells.

2.1.1.3 Exploration

Oil and gas exploration is much more complicated than is typically described. First you cannot open the door and look into Earth directly. You must make measurements from outside—in fact, often through a couple miles of rock. From these measurements, geologists and geophysicists try to deduce where petroleum reservoirs may exist. But the first glimpse inside comes from drilling a well. Even then, direct observations are limited.

The wellbore is long and slender: commonly only six to eight inches (15–20 cm) in diameter at the bottom. The small cuttings from the drilling process are circulated to the surface and observed to understand the rock type and to look for any residual traces of oil and gas. Even if these cuttings show residual oil and gas, it does not prove that commercial quantities exist in the reservoir. So, once the drilling is finished, measurement instruments are lowered into the well and various properties of the geologic formation are recorded that offer clues to what kind of rock is there, how porous is the rock, and how much oil or gas may be contained in the rock's pore spaces. From these data, decisions are generally made as to whether a reservoir contains enough oil and gas to be a commercial success. We are constantly discovering new techniques to gather and to evaluate well data more effectively. Some measurements are now made while drilling the well and new instrumentation designs are constantly being sought to provide more robust information gleaned through the long, narrow wellbore.

Even with the best modern measurements, there are more variables to consider than there are measurements to make in the well. This is, by definition, a numerically underdetermined problem. The analytical approach is, as the saying goes, as much art as science. The art and the science continue to evolve to address more of the complexities seen in petroleum reservoirs. For instance, analytical methods are facing a new evolutionary step to address the very different reservoir characteristics of the shale plays. Many thousands (probably hundreds of thousands) of wells have been drilled and not evaluated with the best existing analytical techniques—either because the wells were drilled before the techniques were developed or they were drilled by operators who did not have access to the most sophisticated analytical techniques. There is certainly a good deal of oil and gas to be "rediscovered" in data from existing wells.[1]

In the early days, petroleum exploration consisted of looking for oil seeps, generally in the form of oil slicks floating on a creek. This search was sometimes coupled with some arcane machinations, such as a "dousing rod," which supposedly dips when passing over the oil reservoir. Although one can still find advertisements for dousing or witching experts, it is difficult to understand the claim at all, considering that even small oil and gas reservoirs commonly extend hundreds or thousands of acres. What would be the point at which the stick would dip? Perhaps walking around with a forked stick was merely an act to draw attention away from the actual task of looking for oil seeps or slicks (which anyone could do if they knew what to look for).

Today the arcane machinations involved in exploration still involves people walking in the field (perhaps after some preliminary work based on satellite imagery). Now they are observing rocks and sedimentary conditions and general geologic history of an area. If the geology appears favorable, then exploration begins in earnest with surface seismic surveys and other geophysical measurements. Surface measurements indicate the general structure of the geologic layers present in the subsurface. Improvements in measurements over the last few decades have allowed increasingly precise geologic cross-sections to be constructed. But this does not indicate where petroleum actually exists. It shows where there are high points in the rock beds, where oil or gas might accumulate.

It is currently impossible to know whether oil and gas are to be found at any location until a well is drilled. Exploratory wells are known as "wildcats." When one is drilled in a region where no petroleum production has ever been established, it's called a "rank wildcat." These are extremely high-risk ventures, with historic success ratios barely better than one in twenty. While some authors like to refer to the advances in geology and geophysics as improving exploratory drilling success ratios, we suggest the improvement is as much a function of focusing on lower risk exploratory drilling (i.e. fewer rank wildcats) during the relatively long period of depressed oil prices from the early 1980s until the beginning of the current century. Once a global shortage pushes energy prices higher—and holds them there—the resulting more aggressive exploratory drilling will probably have more failures again. It will, though, yield some surprisingly large discoveries again (Ebenhack 1995).

International exploratory drilling is often conducted offshore. Contrary to some popular notions, this does not necessarily save the oil companies money compared with drilling on land. Offshore operations often are significantly more expensive. The companies must still spend a great deal of money to acquire the drilling rights and are still obliged to pay substantial royalties to the government in whose waters they operate. The offshore drilling rig and the production platform or subsea completions and pipeline required if commercial oil is discovered are all very expensive. The reality is that oil and gas are found in sedimentary rocks that were originally deposited in the ocean and many of these rocks are still undersea. Another significant factor is that exploration in and around the world is intended to find enough oil or gas to ship to the huge consumer base in the affluent, Developed World. This deters extensive exploration in the interior of Africa, Central Asia, or South America, where extremely expensive pipelines would be required to move oil from the interior to coastal ports. The drilling necessary to determine how much oil and gas reside in the pores of the reservoirs buried under thousands of feet of rock occurs sparingly in remote areas of the world.

[1] This was a large part of Ebenhack's job in corporate headquarters of a major oil company in the 1980s.

Drilling is typically conducted with fluid in the hole that is sufficiently dense that it exerts more pressure inside the wellbore than formation fluids exert. This is, in large part, to prevent blowouts. The infamous Macondo blowout in the Gulf of Mexico (Camilli et al. 2010) occurred after the well had been drilled to its total depth and the dense drilling fluids were replaced with lighter fluids for completion. Since the well is drilled in this "overbalanced" condition, the higher pressure in the well pushes some wellbore fluids into the porous and permeable rocks, pushing the original formation fluids back away from the well. This "drilling mud invasion" complicates evaluations of measurements from the well.

If a well is determined to be commercially viable, then steel casing (pipe) is set and cemented in the well. The casing is then selectively perforated at the depths where oil and gas bearing formations were identified and the oil and/or gas begin to flow into the well. (It appears that one of the mistakes made on the Macondo well that led to the deadly blowout and subsequent environmental disaster was a failure to recognize problems with the cement that was supposed to seal around the casing. The well started to flow at the wrong, indeed worst possible, time.)[2] Virtually all wells start their productive lives flowing.

2.1.1.4 Production

Gas wells commonly flow throughout their productive lives, but the pressure in oil wells eventually declines to the point that they need to be pumped in order to continue producing. The pumping units sometimes referred to as pump jacks, are one of the most recognized images of the oilfield, with a large lever system teeter-tottering back and forth (Figure 2.4). What the pump is doing is lifting the weight of the fluid in the wellbore to reduce the pressure at the bottom so that more fluid can flow out of the formation, before the pump returns to the bottom to lift another load.

Eventually, even with pumping, the well's production declines to the point that it is no longer profitable to produce. Of course, if the price of oil or gas rises, it takes less production to make the same amount of revenue, so it becomes economically attractive to produce each well longer. The price of the oil also determines how aggressively technology may be applied to getting more of the remaining oil.

The hypothetical decline curve of Figure 2.5 illustrates a few points. It was created, assuming a fairly simple field, with four wells, each drilled six months apart. You can see the production rate rising steeply in the first 18 months, when the first three wells are brought on production. Then, it was assumed that each well would produce an initial rate of 100 barrels of oil per day (bbls/day), maintain that production for one year, before beginning a decline, at a fairly common 10% per year. If the field cost $1.5 million per year to operate (including pumping the wells, delivering product to the pipeline, royalties paid to the land owner and, overhead costs), then the economic limit for $50 per barrel of oil would occur at a field production of about 75 barrels per day, shown by the upper arrow in Figure 2.5. With $150 per barrel of oil, the economic limit drops to about 25 barrels per day for the field (the lower arrow). The area between the two arrows is the extra oil that would be produced in this field, merely because higher prices allowed the field to operate longer (almost 10 years longer.) The curve shows how asymmetrical an individual field's production curve is likely to be, as well.

[2] For a compelling look into what all went wrong at Macondo, we recommend John Turely's book *The Simple Truth: BP's Macondo Blowout—A Nonfiction Novel.*

Figure 2.4. Stylized illustration of an oil pump jack in operation. Based on a common industry depiction archived by the American Oil & Gas Historical Society. Illustration courtesy of J. Grannis.

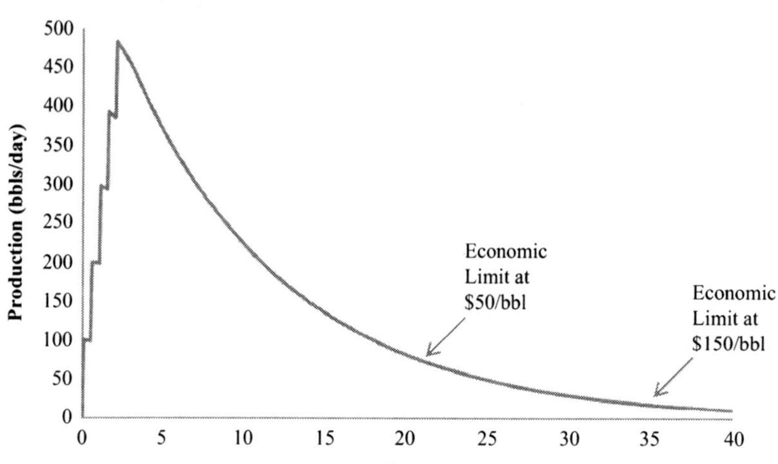

Figure 2.5. Hypothetical oil field production decline curve.

If a well is simply produced until its production rate declines to the economic limit, most of the gas would be recovered, but most of the oil would remain behind. The flow from the reservoir into the wellbore is merely a function of the contents of the reservoir expanding into the lower pressure near the wellbore. This is the driving force to push the oil through the pores to the well. Gas expands very dramatically as pressure declines, while oil does not. It is common to hear this referred to as "sucking" or "squeezing" more oil out of the ground, but neither of these terms is at all realistic.

Consequently, efforts are made to obtain a greater portion of the oil. These technologies are collectively referred to as enhanced oil recovery or simply EOR. EOR typically involves pumping fluids into wells to help mobilize oil and gas left behind and push them toward producing wells in the field. The EOR technique most frequently used is water flooding—injecting water to push the oil. It has been in use for many decades and is still often referred to as "secondary recovery," with more exotic techniques called "tertiary." Even with successful water flooding, about two-thirds of the original oil is likely to be left behind. This is a function of the long, narrow, tortuous flow path through the pores of the reservoir rocks. This is one of the most shocking realities to people unfamiliar with petroleum operations. It is also a major factor in why we can expect petroleum production to decline slowly after a global production peak, as increased prices will support more use of exotic and expensive EOR technologies to extract more of the residual oil. Tertiary techniques, as described later in this chapter, may include chemicals and heat to help move the oil (Ebenhack 1995).

The large share of oil that has been discovered, which remains in reservoirs after they have been produced to their economic limits represents a great deal of potential that can be tapped if prices are higher—and remain high long enough for the technologies to be fully employed. Some experts, though, discount this large potential. As Zittel and Shindler (2002) put it:

> EOR measures have already been applied for more than twenty years, and these measures are accounted for in production forecasts. There will not be any sudden jump in the future—continuous progress is and was always part of the production forecasts. There are two major examples for this: (1) One is the production profile of German oil. After its peak in 1968, the production continuously declined despite efforts to implement enhanced oil recovery techniques. (2) Another example is the production of Prudhoe Bay… This field is at the technological forefront and every possible new measure was exploited to enhance production and to avoid a decline, with almost no success. Today more water is extracted from the wells than oil, water which was injected into the reservoir to increase the pressure.

Some of their observations are summarized below. They suggest that:

1. EOR is only applicable late in the life of a field.
2. EOR is most effective in geologically complex fields, having very low primary recovery factors.
3. EOR is "only effective in the sense that more dollars are gained with the extra oil than have to be spent for the measure."
4. EOR increases the production rate briefly, only to enhance the decline in the long term.

Let's take a closer look at the technologies to understand the real potential and limitations of EOR. Then we can consider some of the specific points raised above.

The first point raised by Zittel and Shindler is intrinsically undeniable. Every oilfield in the world draws on a finite (and plausibly definable) amount of original oil in place. Each well has a finite ability to produce oil, based on the permeability of the reservoir rock and the oil properties. Drilling additional wells boosts a field's production capacity. No EOR technology could hope to cause the field's production rate to increase for any substantial period of time. What EOR can do is to help maintain production levels or to slow the field decline rate. EOR is not a replacement for additional drilling.

Water flooding is the oldest and best-known technique of enhancing oil recovery, which works through the injection of water into some wells to push oil to other producing wells. It was employed in many old fields, with great success. The long-standing rule-of-thumb has been to expect water flooding to double the ultimate recovery that would be achieved without it—going from less than one-fifth of the oil to about one-third. The success of water flooding led to its broad application in many older fields and even some new fields. Other fluids may also be injected, but water flooding dominates the classic secondary recovery technologies.

Thermal recovery is another established EOR technology, which is most applicable to reservoirs with low primary recovery factors (due to fluid, not rock properties). The benefits of steam injection (or more dubious fire flooding) are almost exclusively related to very heavy, viscous oils, which may not flow at all under normal reservoir conditions. Yet another class of EOR is miscible flooding. Here fluids are injected that may allow oil to mix with water (surfactant floods, like using soap to make oil or grease mix with water when washing). This allows more oil to be literally "washed out" of the reservoir with continued injection of water. Or they may inject a gas at high pressure to dissolve into the oil, making the oil flow more easily. For the heavy oil thermal projects and the most promising miscible injection projects, reservoir homogeneity is a greatly preferred condition, quite the opposite of Zittel and Schindler's second point.

Another class of EOR is the polymer-augmented water flood. This would be a case in which reservoir (rock) heterogeneity is being addressed. If there are portions of the reservoir with much higher permeabilities than other portions, any water injected tends to flow into the high permeability section, pushing oil out of it rather effectively, but leaving the lower permeability portions of the field virtually untouched by the water flood. Putting a highly viscous polymer fluid into the wells tends to plug up the high permeability zones, forcing water into the rest of the reservoir.

Finally, the successful application of EOR should yield additional oil, not simply offer a short-term production increase, followed by a steeper decline. The oil production decline on EOR may be steeper, due to the EOR process having formed an oil accumulation, being then pushed along by water. That bank of oil increases oil production sharply, which is likely to lead to a sharper decline. Nevertheless, the oil mobilized and driven to the production wells is additional oil that would not have otherwise been produced. EOR projects are often very expensive and they would not be commercially successful if they did not recover additional oil. Indeed Zittel and Shindler's third point acknowledges that projects may be considered effective merely because "more dollars are gained with the extra oil than have to be spent. . . ." This clearly indicates that extra oil is being produced.

The field production trends that can mislead some are also a function of price. A great many fields that were subjected to exotic EOR techniques in the 1970s were beginning to experience production increases by 1980. Prices fell shortly thereafter and field investments dropped with them. Thus, it is quite possible that the skepticism about EOR is based, at least in part, on a failure to include the role of economics in production practices. EOR will no doubt play and has no doubt played a significant role in moderating the decline rates of many large fields, which will, in turn, moderate regional and global decline rates. On the other hand, it is quite true that EOR will not likely restore fields to their early peak production rates—it will more likely offset their decline rates. It cannot be expected to make possible continued exponential growth in production. Furthermore, the more exotic EOR techniques are not likely to be extensively employed until oil prices get high and stay high for several years. This is likely to come near or after a global oil production peak.

2.1.1.5 Technology Extends the Limits—to a Point

Many people have been wrong about many dire predictions of global shortages of various resources. One of the earlier doomsayers, Thomas Malthus, predicted at the end of the 18th century that the world would soon be unable to produce enough food to provide for growing population levels (Malthus 1798). He failed to foresee the increases in agricultural productivity that were coming. Indeed, the term "Malthusian Predictions" is a popular way of describing unrealistically pessimistic estimates, largely because technology and new ways to tap new kinds of resources have risen to the challenge of providing ever more resources to ever more people. It is perfectly fair to ask why it will not happen again. It is, however, nonsense to imagine that it will continue to happen again and again *ad infinitum*. Because even successful oil fields fail to recover significantly more than one-third of the original oil in place, the EOR technologies we've mentioned have the potential to double reserves in existing fields (one can never expect to recover every last drop, nor to apply the best technologies effectively to every field).

Similarly, new technologies help to find more oil and gas, but there remains an essentially finite amount to be found in the world. Indeed, some of the oil and gas remaining to be tapped is in challenging locations: the arctic, remote jungles, and ultra-deep ocean waters. Of course, efforts to tap the latter have reinvigorated environmental skepticism in the aftermath of the Macondo disaster in the Gulf of Mexico in 2010. There will be many things that are possible, but we may decide not to do them—at least not quickly. We suggest that one lesson of the blowout is that we should not become careless due to overconfidence in our technical ability to do something. Indeed, this will be important for the wide-scale application of many new technologies. Just because we cannot say today what problems may occur with new energy technologies does not mean that problems are impossible or negligible.

2.1.1.6 Untapped Oil and Gas Resources[3]

Oil and gas-producing companies seek to maximize profits. Indeed they have an obligation to shareholders to provide good returns on investments. So, every well must meet commercial thresholds. Those thresholds are remarkably different in the Developed World as opposed to the Developing World. Prospects in locales remote to large, established markets must be about 100 times larger to be commercial than those near markets. For example, the average oil well in the United States produces roughly 10 barrels of oil per day, compared with about 1,000 barrels per day per well in Africa. The disparity is a function of the markets, not of any fundamental differences in the nature of hydrocarbon endowments. Companies see a number of challenges with markets in the Developing World: concerns for political instability, the potential for fields

[3] This section is drawn from a part of a chapter we were invited to write in UNESCO's *International Social Science Journal* (Ebenhack and Martínez in Khodeli 2009).

to be nationalized, and frequently unstable currencies. Therefore, multinational companies only develop oil and gas in the Developing World for export.

When these companies are successful in discovering resources sufficiently large for the demands of the export markets, they bring expatriate workforces to do most of the professional work—and even much of the labor. An expatriate camp is often built, which is practically a small town. Whether or not a separate camp is needed, a large number of other expatriate workers are brought in to provide educational, medical and other support services. Ancillary service companies also move in to log the wells, run casing, cement, and other support. Furthermore, resources tend to be more frequently developed along the coast or offshore, since it will be easier to get the product to international markets than for inland fields. Offshore drilling and production is not cheaper, indeed it is much more costly than onshore, but long pipeline projects can be prohibitive.

There are numerous aspects of operations in a Developing Nation that add to the costs and, thus, the scale at which operations proceed. The cost of bringing a discovery field to production in a new province may be on the order of billions of dollars. All of this means that there is a great deal of overlooked, undeveloped potential, even in regions where exploration has already taken place. There are significant opportunities for reexploration to develop bypassed resources in the Developing World. These resources will be added to the world's reserves. The question will be whether they enter the international, export markets or if they will be used to support local development. Of course, as resources are successfully brought on line, further exploration will be supported, which will also add more reserves.

2.1.1.7 Unconventional Oil and Gas

One of the sources of optimism (which can lead to ignoring the reality of fossil resource limits) is the potential offered by the set of resources commonly referred to as "unconventional oil and gas." As we have discussed, conventional oil and gas resources are those in which the fluid products flow from porous, permeable reservoirs, in which they have accumulated. By contrast, then, the unconventional resources do not naturally flow from porous and permeable reservoirs.

What the industry used to call "tar sands" is now commonly referred to as "oil sands" and has been exploited commercially for several decades (OSTSEIS 2012). Frankly, tar sands made a more descriptive characterization. Many conventional petroleum reservoirs are sandstones, so could be called "oil sands." The distinguishing trademark of this sort of unconventional oil is that petroleum is extremely viscous and dense: tarlike. It is so thick that it cannot flow naturally from the reservoir. There is a continuum of petroleum types from dry gas, to wet gas, to high gas-oil-ratio oil, to low gas-oil-ratio oil, to heavy oil, to tar or oil sands. These "unconventional" resources are merely at the extreme of the continuum. They are classified as unconventional due to the common need to mine the sands in which they occur in order to extract the product. This generally involves a strip-mining type of operation and can incur significant environmental costs.

Nevertheless, these resources have been successfully tapped, especially in Alberta Canada (the Athabasca sands) and Venezuela (Orinoco). There is a tremendous resource base and potential for ultimately recoverable reserves to be larger than those for conventional oil. Still, the environmental externalities and the need for extensive mining or *in situ* processing indicate that the production rates are never likely to be as high as those for conventional oil.

Shale gas and oil also show remarkable potential because these unconventional oil and gas systems exist as fluid oil and natural gas in shales (EIA 2011). In these reservoirs, oil and gas remain in the source rock, simply not having migrated into conventional reservoirs. The fluids are capable of flowing, but the rock has virtually no permeability. Therefore, if fractures are induced in the rock, oil or gas can seep into those fractures and then flow to the wellbore. A great deal of oil and gas has been found to exist in these source-rock shales. This leads to massive hydraulic fracturing. The Marcellus Shale, underlying much of Pennsylvania and southwestern New York has received a great deal of attention, especially relating to concerns over environmental hazards of "frac jobs" (GWPC 2008).

The presence of hydrocarbons in source rocks has been long known, but until recently, of very little interest because of the extremely low permeabilities of the shales. New technologies have caused the industry to take a different look at kinds of rock that would normally have been ignored. Many people think that fracturing—"fracking" in some popular references—is that new technology. In fact fracturing has been a common well treatment for decades and it is being developed and enhanced for these new applications. Probably the most seminal improvement came in "steering" horizontal wellbores. The industry has been drilling wells directionally for decades too, but only around the turn of the 21st century did the ability to control the wellbore telemetry with great accuracy make it possible to drill thousands of feet laterally through a productive zone. This combined with multiple stage "frac jobs"[4] has enabled the industry to go after the vast resources in shales. There is considerable controversy over environmental impacts, but we believe that many of these will be addressed through technology and thoughtful regulatory policies. The need for energy will likely be a great motivator for the cooperation necessary to develop these resources, while mitigating—or at least monitoring—the environmental impacts.

Note, shale gas and shale oil resources should not be confused with *oil shale* resources. What is commonly referred to as oil shale is source rock that does not contain oil, but rather solid, organic material called kerogen. Kerogen can only yield oil upon intense heating, which can be converted into a synthetic crude oil. Thus, oil shale can be viewed as oil and gas source rock, however, the organic matter has not been converted into liquid or gaseous product. Again, unlike shale oil and gas, the oil shales do not contain liquid oil. No substantial commercial production exists, so it's proved, recoverable reserves technically would be zero, in spite of an extremely large resource base. The technique applied in start-up commercial ventures, such as Unocal's *Parachute Creek* project in western Colorado (Duir et al. 1990), has been to mine the shale, crush it, retort it at very high temperature to yield approximately 20 gallons (77 liters) of oil per ton of rock mined. Very little oil from kerogen has been produced from all of the attempts combined. Most current attention to this resource is focused on *in situ* recovery methods, which involve heating the material to very high temperatures while still in the geologic formations and extracting syncrude through wellbores. These are very technologically challenging approaches. The environmental impact would be significantly reduced by successful *in situ* processes, but the energy consumed in producing the synthetic crude will still be very high compared with conventional oil.

It is not clear yet how great the contribution from these unconventional resources will be. Tremendous volumes exist as resource bases for all of them, but the sustained production rates

[4] The industry commonly does abbreviate fracture as "frac"—not "frack." "Frack" was Frick's partner from the old Ice Capades.

are questionable for each. The environmental impacts may also limit development. We believe that the oil and gas produced from shales will contribute significant ultimate reserves. Sufficient engineering could mitigate the environmental problems related to the extensive drilling and the massive fracturing processes. Oil or tar sands have been producing for decades. The technologies are well established and the economics can be favorable. Environmental impact is likely to limit how production expands in the tar sands. Currently, oil shales seem to us unlikely to be major energy contributors, due to the exotic technologies required for *in situ* processes, the large environmental impact of mining and retorting, and the energy inputs required for both technologies.

2.1.1.8 The Future of Oil and Gas

With new geologic discoveries and new technologies, petroleum production rates have grown in response to growth in demand for more than 150 years and will continue to until they are no longer able to do so. When growth in production can no longer keep pace with exponential growth in demand, there will be a peak in petroleum production, followed by a long decline. The only factor that can plausibly impact this reality would be for alternative energy sources to grow and compete effectively with oil and gas before the peak. The improbability of this will be assessed later. Otherwise, the limited resource must reach some limits. But the limits will not be the absolute depletion of the resource, but rather a limitation of *rate.* Even when there is a great deal of oil left to be recovered, it will approach a maximum rate of recovery, as discussed earlier in this chapter.

In fact much literature addresses the popular misperception of a sudden shortage with the expression "peak production" (Hubbert 1982). This is more accurate than the simplistic expression "running out of oil" but is still limited in its commons usage. Surprisingly, even publications by technical experts often fail to explain the controlling phenomena of growth and decline. Production has increased in response to growing demand. Therefore, growth has been a market phenomenon. The peak and ultimate decline, though, will be controlled by physical phenomena. The maximum flow rate from any given well or field is a function of reservoir pressures, permeabilities, saturations, and fluid viscosities. Although market forces will promote more aggressive exploration and production technologies, the decline will be governed by the physical limits within which the technologies must be deployed. This is comparable to the physical limits of human bodies within which medical technologies must operate. Up to a point, medical technology can preserve life and even restore health and vigor, but cannot eliminate the ultimate, invisible limit.

Oil and gas just keeps showing up in new places. This is not to say that we can expect to find another Ghawar—the world's largest oil field, located in Saudi Arabia. Nor is it meant to say that every place geologists dare to dream of will hold vast hydrocarbon riches. In the aggressive exploration of the late 1970s, there was tremendous hope placed on the overthrust belts of the western United States and on a Canadian arctic play called Mukluk, which produced mostly water. Oil and gas remain elusive. An extremely small fraction of the Developing World's sedimentary geologic basins have been extensively explored. A great deal of oil and gas remains to be discovered, especially in the interiors of Developing Nations and in deep waters of the ocean. This will require a great deal of exploratory drilling, including many "dry holes" and some amazing successes.

Uncertainty is one of the key aspects of oil and gas exploration. Geologists cannot "know" what is in unexplored provinces. They actually have to go out and gather extensive information just to see what is exposed at the surface. Are there rocks exposed that seem likely to be associated with the kinds of geologic settings in which oil and gas develop? Are there any natural oil seeps or gas vents that confirm the presence of some oil underground? To understand the subsurface, tests have to be run: generally seismic surveys. Even then, they only give a notion of what the subsurface structures look like, they don't tell where or how much oil and gas exists—drilling is required to confirm that. Indeed, geologic investigation continues to reveal more and more of the many secrets of the earth's complexity. No one can understand all of it, and no one does. Geologists continue to have work, because we continue to need to learn far more about many unexplored regions. You can bet that if the geologists had already totally defined the world's sedimentary basins where oil and gas can occur, the oil companies would not be hiring any more geologists. We'll find a great deal of oil and gas in areas that no one ever thought were likely. Even so, we will face a global peak in production rates in the relatively near term—our rate of consumption will most certainly ensure it. (The peak is discussed in more detail in Chapter 5.)

2.1.2 COAL

Coal, quite literally a fossil fuel, currently represents about one-fourth of global energy supply (IEA 2010). It can form in a number of circumstances, but swamps represent one of the most common origins. The same processes of growth, death, and burial are currently underway in swamps. Indeed, one can travel to some contemporary locations, such as the Dismal Swamps in North Carolina and see potential future coal beds. The swamps represent dense biotic accumulations, as trees and other vegetation die and fall into the stagnant waters.

Although dead plants on land are commonly devoured and undergo decay, the water protects the decaying plant material. Because the water is stagnant, there is very little air at the bottom of the swamp, giving the anaerobic conditions that permit accumulation and decay without oxidation. The carbon remains fixed in place. Even while the swamp is active, the accumulation of decaying organic matter can become a thick layer of peat. In the peat, the anaerobic decay process breaks down complex organic molecules into shorter, simpler ones. Once the swamp is buried under rock sediments, the coalification process begins. The organic material continues to break down into smaller, more carbon-rich molecules, while moisture is driven out and the material becomes denser. Unlike oil or gas, the coal occurs in solid beds, not in the tiny voids of porous rock.

Coal is primarily carbon. As it becomes more pure, it becomes harder, more dense and blacker. Young coals are soft and often called "brown coal" looking rather like the trees from which they are (largely) derived. The oldest coals, the anthracites, are hard and so black that they shine, like obsidian. The more mature coals are, the cleaner and hotter they burn. Anthracite has been a highly prized coal, because it can burn at high enough temperature to undergo very complete combustion, producing very little smoke. This fact reflects the value of the refining process performed by nature in forming all of the fossil fuels. Rather than being naturally dirtier and less efficient, fossil fuels generally offer cleaner and more efficient combustion than the raw biomass from which they are derived. It is the enormity of the consumption enabled by the collective fossil fuels that make their combustion products a serious problem, not an intrinsic inferiority of the fuel. An exception is that fossil fuels may contain small impurities, particularly of sulfur, which produce pollutants. This contamination can be controlled by technology.

2.1.2.1 Coal Distribution

Coal appears to be considerably more abundant than oil and gas. Its resource base is estimated to be approximately four times as large as oil and gas combined. Some of the largest known reserves are in the United States. China is also basing much of its burgeoning development on large coal reserves and looking around the world for opportunities to import coal. Indeed, over 60% of known global coal reserves are in the United States, Russia, and China (see Figure 2.6). More potential is also being developed in the African continent, but the largest known reserves there have been heavily drawn on in the sub-Saharan nation of South Africa.

Similar to oil and gas, there are numerous unexplored regions in the world. New coal fields will be found and developed throughout the world for decades, perhaps even centuries, to come. (If not for domestic electricity production, then most likely for industrial purposes.)

2.1.2.2 Coal Exploration and Production

Being a solid fuel, coal is conventionally obtained by mining. The earliest forms of coal mines were "drift" mines into the side of a hill (where erosion had exposed part of the coal seam). It started out as merely digging back into the hillside as far as one could reach. But the value of the energy the coal produced provided incentive to try to get more of it out of the ground.

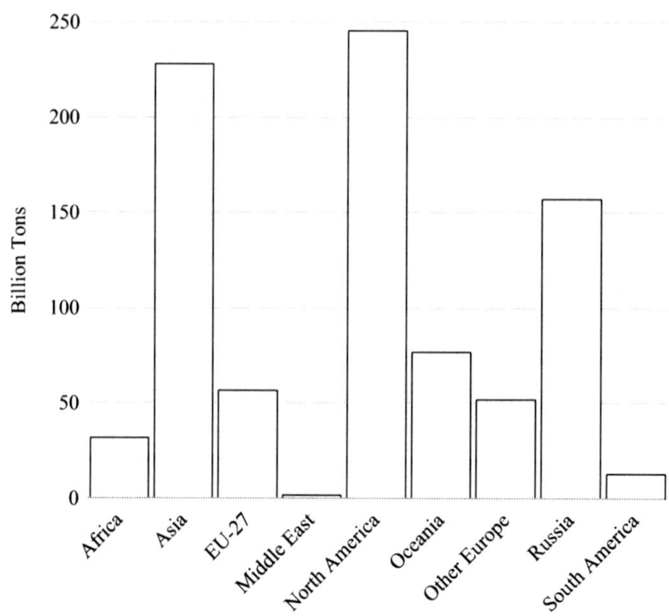

Figure 2.6. Proved recoverable coal reserves by region, end 2008.
Source: WEC (2010).

Learning from many small disasters as early drift miners were buried under roof collapses, more technology was brought to bear. Log timbers, rock, or even brick were used to shore up the mine roof. Light was needed to see and explosions from the coal dust and from methane gas were common in those early years, when open-flame lanterns were the only way to bring light into the recesses of the mines. This led to some ingenious engineering to make lanterns that drew air through a small pool of water to prevent the flame from being in direct contact with the gasses in the mine. Underground mining progressed to sinking shafts to reach deeper coal beds and then extracting the coal in "room and pillar" mining. The pillars (more like walls) are coal that was left behind to support the roof. The pillars might comprise some 60% of the total coal bed. Of course, then, the value of the product led to efforts to "rob the pillars" at the end of a mining operation, fully expecting the roof to collapse behind it (Ebenhack 1995; from Moore 1940).

Coal mining has a rich history as extraordinarily hazardous work. If the miner survived the threats of explosion, roof collapse or flooding, he was at great risk of the long-term health hazards, collectively referred to as "black lung disease." Technology has done a great deal to mitigate these hazards, but underground coal mining remains one of the most dangerous civilian occupations. One of the technologic advances came in the development of enormous machinery for surface mining (or open pit or "strip" mining). In this process, the overlying soil and rock are stripped away by the massive machinery, exposing the coal bed to be extracted in its entirety. Strip mining has been one of the most visible and contentious forms of energy acquisition. Even passersby have seen some of the huge open pit mine and been appalled by the massive disruption of the terrain and local ecosystems. In 1977, the United States Surface Mining Control and Reclamation act required that mine operators pay into a fund and prepare plans for reclamation and restoration of the surface contours.

While strip mining may tear 50 feet (15 meters) or more of the earth away, the coal seam itself is likely only to be a few feet thick. Piles of these so-called "spoils" are pushed back into the hole after all that has been extracted. Graders and bulldozers now work at restoring the original contours, as well as ensuring that the rock that came out last goes back in first and that topsoil is restored to the surface. Regulations also now require planting vegetation on the restored topsoil, to help mitigate erosion. The short-term disruption is still immense, but current operations in the United States and much of the Industrialized World require prudent operations with a plan to reclaim the land. In passing by you may recognize former strip-mining areas by the stands of young pine trees. The evergreens are fast growing, so they are planted in the reclamation process to provide vegetative cover and establish root systems to prevent erosion. Where pine forests are not native to the ecosystem, one can readily see the native hardwoods reestablishing themselves and mixing with the pines within a few decades.

The scale of the coal mining operations is massive. Consider the Big Muskie used in east central Ohio in order to strip the overburden in open pit mines to get down to the coal (Ebenhack 1995). The bucket it dragged along a 310-foot boom, scooped up 220 cubic yards in a single load—the equivalent of a two-story home. It was operated by 170 electric motors, each producing anywhere from 40–2,000 horsepower. Within the mine, trucks carry 200 tons—that's twice as much as a railroad car. Today the Big Muskie is retired, but the scale has not really diminished, with mountaintop removal becoming a common process especially in the Appalachian region of the United States. As the name suggests, this involves removing the top of a mountain, placing the remnants in the neighboring valley, in order to expose a coal seam. This is a highly controversial technology, since it does disrupt the topography and local ecosystems substantially.

2.2 NUCLEAR FUELS

There are two fundamental kinds of nuclear power: fission and fusion. Both of them are based on converting small quantities of mass to energy, following Einstein's famous equation $E = mc^2$. E represents energy and m represents mass. Since c represents the speed of light, which is an astronomically large number, multiplying by the square of its number represents a vast amount of energy derived from a miniscule conversion of mass. If a process could be developed that took advantage of this principle and captured the heat released by the "splitting" of some atomic fuel used in this mass-to-energy conversion, it can then be used to generate electrical power in a fashion quite similar to other combustion processes that use fossil fuels. As of yet, only fission has been proved to be a viable energy source.

2.2.1 FISSION

Nuclear fission involves splitting the nuclei of heavy atoms of primarily uranium. The naturally occurring fissile isotope U-235 undergoes fission when an additional neutron enters the nucleus (see Figure 2.7). This nuclear reaction releases approximately 195 million electron volts per atom—compare this to a meager 30 electron volts of energy per molecule of exploded TNT and you can see the power of nuclear fission. The fission process yields two smaller nuclei (lighter elements) and several high-energy neutrons. These neutrons strike more fissile nuclei, creating a chain reaction of ongoing fission, yielding more high-energy neutrons and more fission (Vandenbosch and Huizenga 1973).

A great deal of elemental uranium is spread throughout the earth's crust, but in very small concentrations. Beds containing more than 0.2% uranium are commonly mined, with 1% concentrations making for extraordinarily rich beds. Of the uranium which is extracted, only 0.8% is the desirable U-235 used in conventional fission reactors. The separation of sufficiently pure U-235 is an essential step that involves very sophisticated, expensive technology and some energy input. Lower grade occurrences of uranium ores certainly exist, but require more costly processing. Even though the energy yield from fission is very high, the limits to viable uranium sources ultimately set constraints for the future.

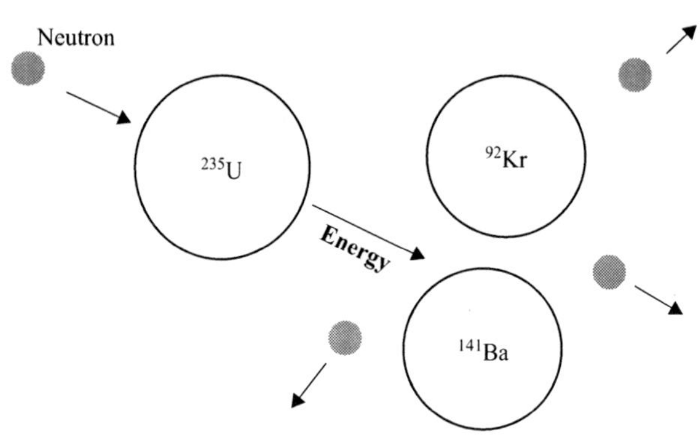

Figure 2.7. U-235 nuclear fission reaction.

There is, however, a technology, called breeder reactors, which makes use of some of the neutrons produced in the nuclear chain reaction to bombard the nonfissile uranium nuclei. If the neutrons are slowed down, some are absorbed in the nuclei of nonfissile, but fertile isotopes. The absorbed neutron makes the nucleus unstable, which causes radioactive decay, which, in turn, converts the fertile isotope into a fissile one. For example, when fertile U-238 absorbs a neutron, it undergoes prompt release of two beta particles (effectively electrons separating from the neutron turning it into a proton) it ends up as fissile Pu-239. Thus a breeder reactor, breeds more fuel from what was previously nonfuel (nonfissile) material. This technology could increase the available fission resource base by more than two orders of magnitude, since there is less than 1% U-235 naturally occurring, but more than 99% fertile U-238. The technology has worked in a series of French reactors, operating from 1984, until the end of 1996, when the largest reactor, known as the Superphénix, ceased producing electricity due to safety reasons (IAEA 2010).

2.2.2 *FUSION*

Nuclear fusion is a popular topic in discussions of alternative energy sources, because the fusion reaction is even much more powerful than fission. It is based on forcing nuclei of small atoms together. In the process, a small amount of mass is transformed into energy, but slightly more mass per atom than in fission reactions. The sun is a giant reactor, fusing isotopes of hydrogen to form helium. (Larger stars bring together larger atomic nuclei, making heavier elements, but our sun only makes helium: the same as all of the research on controlled fusion.) Because fusion requires forcing small nuclei together at high energy, unlike fission, the fusion reaction requires substantial energy input to initiate. Since no controlled fusion reaction has yet been able to produce a sustained net energy gain, its reserves are exactly zero, while its resource base is immense. It has the potential to be a game changer for global energy systems, but this will require some substantial technical breakthroughs to be achieved (Miyamoto 2005).

The central challenge is to bring nuclei sufficiently close together at sufficiently high energy for sufficient time to support fusion reactions. These requirements work against each other. The nuclei tend to be positively charged and thus repel each other. Higher energy is seen in the form of higher temperature, which tends to cause expansion, which also drives particles farther apart. No earthly materials would be capable of physically holding the fusion material at the millions of degrees at which the nuclei would fuse. There are two fundamentally different technologies competing for the prize of controlled fusion energy production. One is to use powerful magnetic fields to contain the fusion material and reactions. The magnetic fields would be relatively unaffected by the immense heat. The Tokomak is one version of this, in which a toroidal (doughnut-shaped) vessel is surrounded by powerful magnetic coils and the fusion material, in plasma state circulates at extremely high energy around the torus. Two major Tokomak tests are being developed at Princeton and in the United Kingdom (PPPL 2012; CCFE 2012). The other major technology is called inertial confinement. It can be thought of as employing the principles of nuclear fusion weapons on a much smaller scale. A tiny sphere of nuclear fusion fuel (the isotopes of hydrogen, deuterium, and tritium) is dropped into the confluence of numerous powerful lasers to impart the threshold energy, before the material flies apart. Neither technology has achieved a net energy gain yet, over the amount required to initiate the reaction (Boyle et al. 2003).

2.2.3 URANIUM DISTRIBUTION

Uranium resources to provide fuel for nuclear fission reactors are scattered in sedimentary and metamorphic rocks around the world, with large shares of the reserves having been developed in Australia, Russia and Kazakhstan, the western United States and Canada, and southern Africa (WEC 2010). Since nuclear fuel can be enriched to make nuclear weapons, there are nuclear proliferation issues with both the technology and the development of uranium resources. If nuclear fusion technology becomes a viable energy producer, the deuterium and tritium isotopes of elemental hydrogen needed to carry out the reaction will be very easy to find. Since every molecule of water has two hydrogen atoms, there is an abundance of deuterium and tritium in the world. This abundance is what fuels the quest for controlled fusion. However, it should be noted that, as hydrogen isotopes are extracted from water, they become slightly less abundant, making future extractions harder. This would not be a factor for a very long time, though, since so little is needed to produce vast quantities of energy.

2.2.4 URANIUM EXPLORATION AND PRODUCTION

Exploration for uranium is rather similar to that for fossil fuels: starting with identifying likely geologic provinces and searching for evidence of accumulations of the radioactive ore. Test holes may be drilled to seek the underground deposits. A little more than half of the world's uranium ore production currently comes from conventional underground or surface pit mines, similar to those used for coal. Another 36% comes from *in situ* leaching. In the latter process, wellbores are drilled, like those for oil and gas production. Acidic fluid is pumped into the well to dissolve uranium salts and then pumped back to the surface and the uranium extracted. *In situ* leaching is increasing its share of total uranium production (World Nuclear Association 2011). Underground mining hazards for uranium are comparable to those for coal, but because nuclear energy is so much more intense, far less fuel needs to be mined to make the same amount of energy, thus fewer workers are exposed to the hazards.

Once uranium ore is mined, the work has really just begun. After the uranium is separated from the other minerals in the ore body, it is necessary to separate U-235 from the nonfissile U-238. This cannot be done with chemical processes, since the various isotopes are chemically identical. The physical difference is mass per atomic unit. The fissile U-235 is very slightly lighter (about 1.26%) than the more abundant U-238. So, the uranium is reacted with hydrofluoric acid to make uranium hexafluoride gas, which is put into a centrifugal separator and the more dense molecules are driven to the outside and removed, leaving enriched U-235 behind—but the fuel still only contains a little more than 3% fissile U-235, but that's enough. Finally, this enriched fuel is formed into rods, which are bundled together for use in the reactor power plant core (Kraushaar and Ristinen 1993).

Since the fuel source for fusion would be hydrogen, from water, no real exploration process is necessary. Deuterium and tritium are isotopes of hydrogen that can be found wherever there is hydrogen, such as in water. The technology for separating the deuterium and tritium isotopes from the rest of the hydrogen will be similar to that used to separate U235 from U-238, based on the slight difference in mass between the isotopes. In the case of hydrogen, the lightest element in the universe, though, the difference is relatively much larger (approximately 100%).

REFERENCES

Boyle, G., B. Everett, and J. Ramage. 2003. *Energy Systems and Sustainability*. New York: Oxford University Press.

Camilli, R., C. M. Reddy, D. R. Yoerger, B. A. S. Van Mooy, M. V. Jakuba, J. C. Kinsey, C. P. McIntyre, S. P. Sylva, and J. V. Maloney. 2010. Tracking hydrocarbon plume transport and biodegradation at Deepwater Horizon. *Science. 330* (6001), 201–204. doi: http://dx.doi.org/10.1126/science.1195223

CCFE (Culham Centre for Fusion Energy). 2012. http://www.ccfe.ac.uk/

Duir, J. H., A. C. Randle, and C. P. Reeg. 1990. Unocal's parachute creek oil shale project. *Fuel Processing Technology. 25* (2): 101–117. doi: http://dx.doi.org/10.1016/0378-3820(90)90099-E

Ebenhack, B. 1995. *Energy Resources: Availability, Use and Impact.* Tulsa: PennWell Publishing.

EIA (U.S. Energy Information Administration). 2011. *Review of Emerging Resources: U.S. Shale Gas and Shale Oil Plays.* http://www.eia.gov/analysis/studies/usshalegas/

E&P (Exploration and Production Magazine). 2009. "Tullow Oil announces major oil discovery in Uganda." http://www.epmag.com/Production-Drilling/Tullow-Oil-announces-major-oil-discovery-Uganda_27598

GWPC (Ground Water Protection Council). 2008. *Hydraulic Fracturing Considerations for Natural Gas Wells of the Marcellus Shale.* http://www.dec.ny.gov/docs/materials_minerals_pdf/GWPCMarcellus.pdf

Hubbert, M. K. 1982. In, *Perspectives on Energy: Issues, Ideas, and Environmental Dilemmas.* (Ruedisili, L. C. and M. W. Firebaugh, eds) New York: Oxford University Press.

IAEA (International Atomic Energy Agency). 2010. *Fast Reactors and Accelerator Driven Systems Knowledge Base: Phènix—30 Years of History: The Heart of a Reactor (Jean-François Sauvage).* http://www.iaea.org/inisnkm/nkm/aws/fnss/phenix/book/index.html

IEA (International Energy Agency). 2010. Key World Energy Statistics. http://www.iea.org/textbase/nppdf/free/2010/key_stats_2010.pdf

Khodeli, I., ed. 2009. *From Curse to Blessing? Using Natural Resources to Fuel Sustainable Development.* Hoboken: Wiley-Blackwell.

Kraushaar, J., and R. Ristenen. 1993. *Energy and Problems of a Technical Society.* New York: Wiley.

Malthus, T. R. 1798. *An Essay on the Principle of Population.* London: J. Johnson.

Miyamoto, K. 2005. *Plasma Physics and Controlled Nuclear Fusion.* Berlin: Springer.

Moore, E. S. 1940. *Coal.* New York: Wiley.

OSTSEIS (The Oil Shale and Tar Sands Programmatic Environmental Impact Statement). 2012. *Oil Shale/Tar Sands Guide.* http://ostseis.anl.gov/guide/index.cfm

PPPL (Princeton Plasma Physics Laboratory). 2012. http://nstx.pppl.gov/overview.html

Schlumberger Oilfield Glossary. 2011. http://www.glossary.oilfield.slb.com/

Vandenbosch, R., and J. R. Huizenga. 1973. *Nuclear Fission.* San Diego: Academic Press.

World Energy Council. 2010. *Survey of Energy Resources 2010.* http://www.worldenergy.org/publications/3040.asp

World Nuclear Association. 2011. "World Uranium Mining." http://world-nuclear.org/info/inf23.html

Yergin, D. 1991. *The Prize: The Epic Quest for Money, Oil and Power.* New York: Simon and Schuster.

Zittel, W., and J. Schindler. 2002. "Future World Oil Supply." International Summer School on the Politics and Economics of Renewable Energy. Salzburg, Germany. http://www.peakoil.net/Publications/International-Summer-School_Salzburg_2002.pdf

CHAPTER 3

"Renewable" Energy Resources

This chapter discusses solar, solar-derived, tidal, and geothermal resources, the resources that comprise the so-called renewable energy supply.

3.1 A NOTE

Before we delve into the content of this chapter, we feel it is necessary to begin with a reminder regarding the term "renewable" and to explain the general nature of the chapter, especially with respect to the previous chapter on nonrenewable energy. In Chapter 1, we discussed in detail the problems that come with using terms like "renewable" and "nonrenewable" to define an energy resource. They are incomplete in the way that they are used because they do not always include the way a particular resource is consumed with respect to the way it is, or can be, replenished. A perfect example is the sun. It is an aging star, and in time will reach the end of its life, well after becoming unfit for sustaining life on Earth. But since that process will take billions of years, we consider the energy that Earth receives from the sun to be "nondepletable," hence the usage of the term renewable. This includes the direct and indirect forms that will be described in the material below. Of course, as also was discussed in Chapter 1, it is possible to deplete readily available and renewable energy stocks that were derived from solar energy (e.g., biomass and fossil fuel) if the rate of consumption exceeds the renewal rate. And consumption often does exceed renewal for most of the energy sources that humanity harnesses—fossil fuels by several orders of magnitude. Likewise, geothermal and tidal energy may be considered renewable resources because of the processes by which they are replenished, but also are limited by their net rate of renewal, if accounting for consumption.

Furthermore, it is not practical at the writing of this book to discuss established operational norms for most of the renewable energy systems in the same way as was done for the fossil fuels in Chapter 2. Oil, gas, and coal are major industries, served by established specific professions, such as petroleum engineering, mining engineering, and geology. While the fossil fuel industries are not as homogeneous as many people seem to believe, the professions that have grown up to serve them do establish a basis for operating procedures and best practices. Of the renewable systems, only large-scale, dam-based hydropower is comparably established, perhaps along with traditional biomass energy used for process heat and electricity in certain regions. In the past, wind power was well-established, being sustained for centuries as a source of mechanical energy for pumping and grinding. The newer industries growing around wind energy and solar energy have not yet developed such large-scale, global operational norms, though they are now maturing rapidly. Indeed, successful efforts will emerge and develop their own sets of norms and (probably) engineering disciplines specific to these fields, but that is still burgeoning. Just as much of the work on "new renewables" remains in the realm of basic science, research, and pilot projects, our discussion in this chapter will focus largely on the basic principles behind these emerging technologies.

3.2 EARTH'S ENERGY ALLOWANCE

All of the energy resources on Earth can ultimately be traced back to nuclear processes and gravity. Indeed, the sun can be viewed as a vast thermonuclear fusion reactor, generating tremendous quantities of heat and light that are radiated out to space in all directions. Most of Earth's total energy allowance results from direct solar radiation, able to be used immediately, either for heating or electrical power. But solar radiation also powers other continuous Earth-bound processes such as: (1) the evaporation and precipitation of water in the hydrologic cycle; (2) the atmospheric movement of air due (in part) to uneven warming of Earth's surface—known as wind; and (3) the creation of biomass through photosynthesis. These processes are the origins of what we know as hydropower, wind power, and biofuels, including fossil fuels, which all are indirectly utilized solar radiation. In addition, the nuclear decay of naturally occurring radioactive isotopes in Earth's interior is responsible for maintaining geothermal heat. Finally, tidal forces, unrelated to nuclear processes, are primarily a result of the gravitational attractions between Earth and its moon.

Due to its connection with the sun then, Earth has both a continuous influx of energy, often referred to as an energy income, as well as a storehouse, or "stock" of energy, which similarly can be referred to as an energy savings account. The income portion represents a relatively sustainable total energy budget, which is composed of the constant bombardment of radiant energy entering our atmosphere from the sun, plus tidal gravitational energy due to our immediate relationship with the moon, plus geothermal energy being produced primarily by Earth's radioactive core, minus any energy that is reflected directly or reradiated out to space. Within the solar contribution that penetrates the atmosphere, there is direct conversion of radiation into heat, with the remainder becoming indirect energy that contributes to the hydrologic cycle, photosynthesis, and wind and waves. Meanwhile, the "savings account" of stored solar energy includes fossil fuels, and biomass stocks, especially in forests. Thus, continuous energy available on Earth can be reduced to direct solar energy, indirect solar energy, geothermal energy, and tidal energy.

3.3 THE SOLAR RESOURCE

Our sun is classified (spectrally) as a G2 V dwarf star (nasa.gov). The G2 designation refers to the star's surface temperature, approximately 5,800 Kelvin. The V designation refers to its method of energy generation, which is the fusion of hydrogen nuclei into helium.[1] The sun's central thermonuclear reactor is converting massive amounts of matter into energy

[1] Fusion research, described very briefly in Chapter 2, aims figuratively to "put the sun in a bottle."

every second. This radiant energy is able to travel relatively unobstructed through space to other objects, including Earth's atmosphere. This constant bombardment equates to roughly 1,370 watts per square meter of flat surface. Earth being a sphere reduces this number to about 340 watts per square meter, which is the solar radiation incident on Earth's atmosphere. Reflection from clouds, air, ice, and water plus absorption by the atmosphere and by clouds further reduces this constant to about 170 watts per square meter incident on Earth's surface (Figure 3.1). As per Smil (2008) and others, this amounts to about 87 petawatts (10^{15} watts) of power hitting Earth on average—ample energy to satisfy the appetites of our world's populations.

Most modern anthropogenic uses of incident solar energy require trapping the energy and converting it into electricity or heat. The technologies available to convert solar energy into useful heating and into electricity are still relatively inefficient, due to a number of quantifiable factors. One important factor is that solar energy is actually very diffuse, causing it to have a very low point-specific power density. This means the sun does not deliver very much energy to any one location at any given time, thus requiring the need for large surface areas in order to collect useful quantities of solar energy. Another important factor is that while solar energy hits the atmosphere at a constant rate, it does not hit the *surface* at a constant rate. Indeed it varies depending on the location, the time of the day and year, and local weather conditions. Thus, despite the vast solar energy potential, there are a number of factors that prohibit humanity from tapping the solar resource adequately to power society in a fashion similar to the current system (i.e., massive stock-based energy consumption). Even so, the promise of a solar powered society is too attractive to let go, thus, the on-going human pursuit of harnessing the resource. Three established technologies that are continually researched to improve efficiency and energy output are described below.

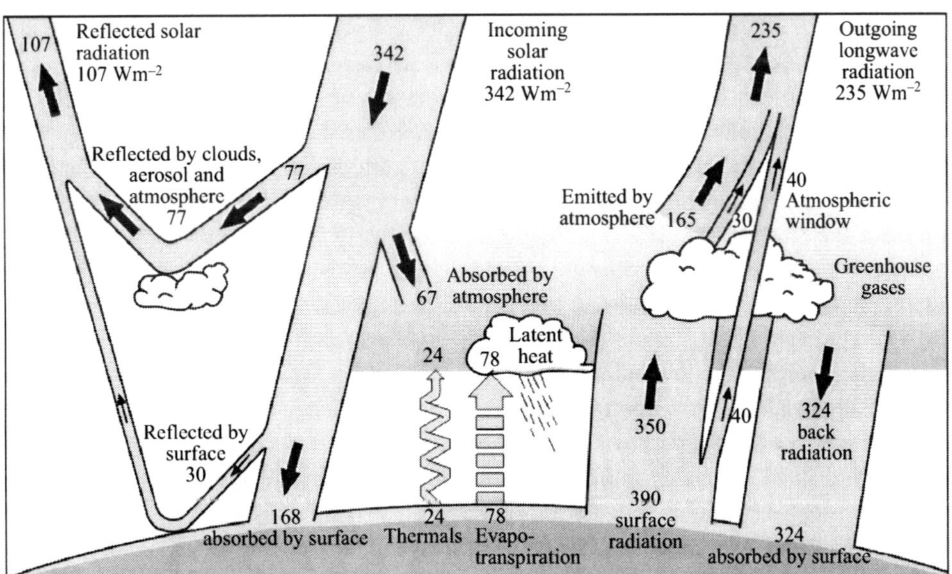

Figure 3.1. Earth's solar radiation balance.
Source: IPCC (2007) adapted from Kiehl and Trenberth (1997) © Copyright [1997] American Meteorological Society (AMS).

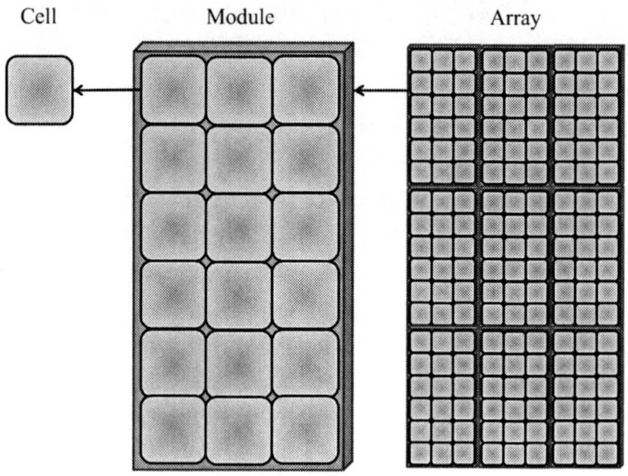

Figure 3.2. A solar PV cell, module, and array. The larger the array, the more electricity can be produced.

3.3.1 SOLAR PHOTOVOLTAIC TECHNOLOGY

When most people hear "solar," they commonly think of solar photovoltaic (PV) technology. PV has become ubiquitous for small-scale, portable technologies, such as in calculators and small battery chargers for phones and other electronics. Photovoltaic energy itself is derived by converting light—photons—directly into electricity. When a photon strikes an electron, the electron becomes excited and jumps out of its orbital. In certain semi-conducting materials, like silicon[2], the electron jumps into what is known as a conduction band, which can be exploited to create an electric current. With a little creative design, the semiconductor material can be fabricated in a way that some negative "N-type" donor sites and some positive "P-type" receiver sites are layered into the material. An individual PV cell, with its own positive and negative layers, creates an electric field and the electrons that are freed by incoming solar energy generate an electric current, which can be harnessed by devices that are powered by electricity (see Chiras 2009). If PV cells are linked together in a module then this electricity is multiplied by the number of cells housed in the module. And if modules are placed in an array, then, depending on the location, there may be sufficient electricity to help power buildings, with the potential to tie into local electric power grids (see Figure 3.2).

Energy and materials (including potentially hazardous materials) are required in the manufacturing of PV cells. However, once a solar PV cell is built and connected into an electrical circuit, it will produce "clean," emission-free electricity whenever it is struck with sufficient light. Many authors refer to the low efficiency of PV cells as a big obstacle to wide-scale adoption: commercially available units operate at about 10–15% efficiency (Boyle 2004). But thinking about this more critically, we know that a typical car achieves about the same net efficiency in converting chemical energy into work—and we all know how many people all across the world have adopted the automobile for transportation. The problem is much better described by the low intensity of solar power and its intermittence and unreliability based on local weather

[2] Other materials are being explored, such as a combination of copper, indium, gallium, and selenium, which may have some important near term potential for inexpensive thin film technologies.

conditions. It does not produce "on demand," and it takes a great deal of area to collect much solar power. Increased efficiency can help to offset the low intensity problem, but does nothing to address the on-demand production failings. Thus, solar PV must be coupled with other energy sources or with storage technologies, such as batteries, flywheels, and compressed air, among others.

3.3.2 CONCENTRATING SOLAR POWER

Solar power also can be harnessed to produce electricity by concentrating sunlight onto a singular location. This is achieved by reflecting it against huge mirrors, then focusing it onto a point or a line, and then boiling a working fluid to drive a traditional turbine power system to generate electricity. In principle, this is a simple and efficient design, with the largest limiting factor being available sunlight. There are many working power plants and several more pilot installations around the world operating at the scale of several megawatts (MW) and several more under development at a full commercial scale of a few hundred megawatts (St. John 2009). The thermal mass of the working fluids actually creates some intrinsic storage to help mitigate the intermittent nature of the solar source. There are many types of these concentrating solar power (CSP) technologies, including towers, dishes, linear mirrors, and parabolic troughs: the most successful CSP technology (Figure 3.3).

CSP is likely to continue for some time to be the most successful solar technology for centralized power plants, because of the large possible electrical output and because of the thermal storage capability. However, even parabolic trough technology—currently the lowest cost CSP option for producing electricity—still requires heavy subsidies to be competitive with traditional power plants. Currently, unsubsidized electricity from troughs costs about twice that from conventional sources in the United States (EERE 2012). Of course, the larger the scale, the more cost competitive the technology can become. Indeed, some people have drawn the conclusion that concentrated solar thermal electric power can only be accomplished on a large scale. However, there is no fundamental physical reason that this should be so. A group at MIT rose to the challenge, developing a microsolar thermal application, particularly for the Developing World (MIT 2006). This remains an economic challenge more than a technological one.

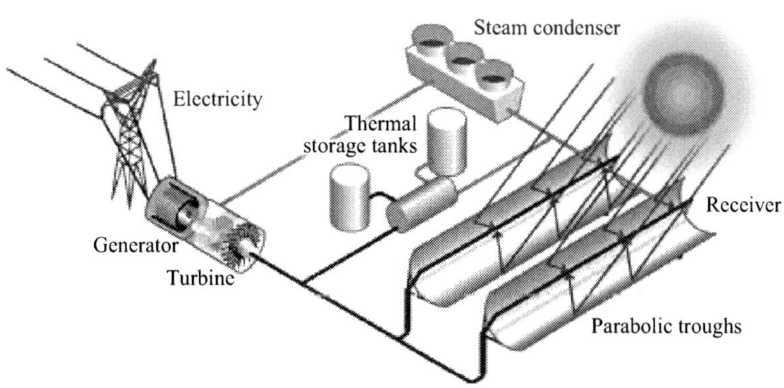

Figure 3.3. Horizontal concentrating solar power system (also known as a parabolic trough system).
Source: EERE (2012).

Another solar technology that is appropriate for small-scale applications is the solar pond. This can be viewed as a low-cost, low-tech CSP. Here, a broad, but shallow, pond is filled with water layers of varying salinity. The deepest layer is the most saline, with shallower layers being successively fresher. The salinity difference prevents natural convection, because the more saline water is sufficiently denser than the next layer above it, such that the natural expansion with heating does not make it lighter than the overlying layer. Thus, as the sun shines onto the dark bottom of the shallow pond, the deepest layer can become quite hot, without the water carrying the accumulated heat to the surface. This temperature difference can be tapped to generate electricity. Even though the difference is much smaller than in the types of CSP described above, with resulting lower efficiencies, the cost is also much lower (Twidell and Weir 2006). It would have no real potential for large-scale, centralized power plants, but might have potential for remote and Developing Nation applications.

3.3.3 PASSIVE SOLAR ENERGY

Passive solar energy collection is quite simple: place objects (i.e., buildings or collectors) in strategic locations to receive natural light and heat so that you do not have to light and heat by artificial means. Passive solar designs that utilize "day lighting" have been employed since the first use of windows. This technique combines both type and placement of a window with interior building design to control how sunlight enters, with advantages for heating, cooling, and internal lighting (DOE 2011). Window types vary widely, with modern technologies and designs allowing for multiple layers of glass, separated by air or an even less conductive gas, like argon, so that passive solar energy can be tapped more effectively. In the northern hemisphere, thoughtful architectural designs place windows to face south in order to admit the most possible light in the winter months, with the least direct sun during the summer to reduce the need for artificial cooling (Figure 3.4).

Passive solar design can also use eaves that extend sufficiently far out to block most of the direct overhead sun in the summer, when heating is not desired, and to have dark masses

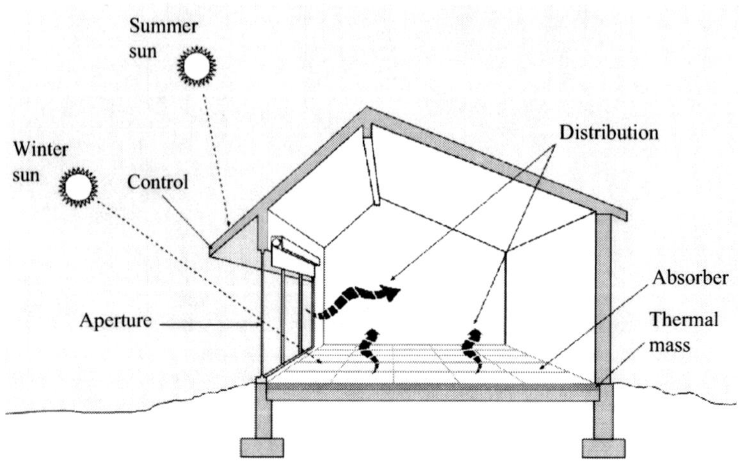

Figure 3.4. Passive solar design.
Source: NREL (2001).

positioned for the sun to warm. The technology can include designing openings at the top and bottoms of walls being warmed by the sun, so as to optimize the natural convective airflow. In order to store the heat through a cold winter's night phase change materials can be employed that change from solid to liquid at about room temperature. The phase change stores a great deal of heat, which slowly releases into the surroundings at night, to maintain a more consistent temperature, as the substances returns gradually to the solid phase as they cool.

Based on similar principles, solar collectors have been devised to heat water for washing and bathing. The technology is relatively straightforward and largely a plumbing project. The basic idea is that if water resides in dark pipes exposed to direct sunlight, it will warm. (It is quite easy to observe the phenomenon, even with a garden hose laying out on a sunny summer day.) This phenomenon is further promoted by housing the plumbing in a box with a transparent cover and dark interior. Sunlight passes through the transparent cover and is largely absorbed and converted to thermal energy by the dark mass of the pipe and container. The only other requirement is to move the water through the system, which may require a pump. Additionally, an antifreeze fluid must be used in regions where outside temperatures fall below the freezing point of water so as not to damage the system, especially at night (NREL 2003).

In most regions of the United States and northern Europe, insolation (sunshine reaching a point on the surface) cannot provide all of the energy for heating water, as there may be prolonged periods with little sunshine. This problem can be mitigated by having very well-insulated hot water storage tanks, but some on-demand heating is commonly necessary to augment the heating during periods of limited sunshine. Nevertheless, solar water heating has the potential to make a noticeable impact on domestic/residential energy consumption: perhaps several percent of that sector's demand.

We want to emphasize here that passive solar energy is perhaps the most readily utilizable resource humanity has, but, ironically, is perhaps the most underutilized. With the availability of cheap and abundant fossil energy, design for natural lighting, heating, and cooling has often been neglected with homes being built for convenience and aesthetics, disregarding solar harvesting potential. Any heating and lighting needs are then commonly met by burning fuels: often on the spot, as with kerosene and candles, or at power plants to generate electricity. Given the resource and environmental constraints that are emerging in the 21st century, design and behavior changes will be required to move away from 20th century energy use norms.

3.3.4 SOLAR ENERGY DISTRIBUTION AND INSTALLED CAPACITY

Every part of Earth receives an average of about 12 hours of sunshine per day. Near the equator, there is relatively small variability in this amount, while near the poles, the sun never quite sets around the summer solstice and never quite rises at the winter solstice. But the sunshine is never very direct in the high latitudes near the poles. Insolation is also significantly blocked or diffused by cloud cover, reducing the potential for solar power production at the surface.

The ideal locations for solar power generation then, are at relatively low latitudes in relatively arid climates (see Figure 3.5). In many places, even if there is a protracted rainy season, though, there may be a few hours of sunlight most days, making solar power viable. For all of the solar technologies, a limiting factor is the intermittent and unreliable nature of insolation. It is intermittent on the predictable cycle of day and night and seasons, while being unreliable due to weather.

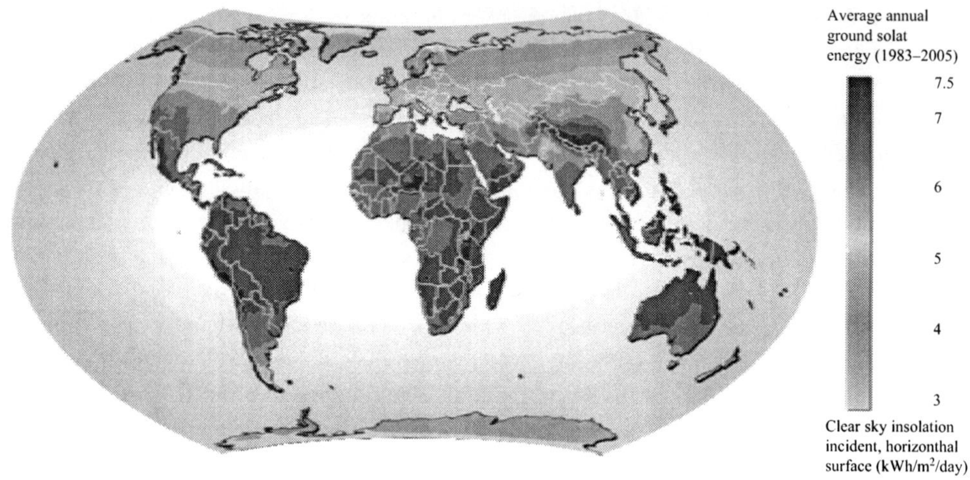

Figure 3.5. Clear sky ground solar energy distribution.
Source: Courtesy of UNEP/GRID-Arendal.

These factors dictate that energy storage is an important part of stand-alone solar power projects. Battery back-up is most commonly employed and it is not unusual for the cost of batteries to be approximately equal to the cost of the actual energy production. This represents a significant economic deterrent for stand-alone systems. CSP projects can have some advantage in this regard, as there is some innate storage in the heat built up in the fluids and working system, as opposed to PV systems, which cease producing the instant that light ceases to strike the cell.

Solar PV technology tends to be the most effective for relatively small power demands in remote locations where price is not a major constraint. An early adopter of this technology was the space program, where all of the criteria were met—and once leaving Earth's surface, the weather and daylight issues were eliminated. Applications with lower power densities that are mobile or that are sufficiently remote from grid power are increasingly adopting PV technology. In some developing country locations, it is found that grid-generated electricity is sufficiently expensive and/or unreliable that solar PV makes current economic sense. The applicability will increase as energy prices rise, but it remains unclear how PV will contribute in the broader market for generating electricity for main power grids.

Quantification of the total potential of the solar resource can be an interesting academic exercise and useful for future planning of solar-based projects. The current capacity that actually exists to generate energy or electricity is extremely small. Given the variable nature of the solar resource and the relatively high capital cost to install solar technologies, compared to historically cheap fuels, increasing capacity is primarily tied to government incentives. The feed-in tariff seems to rule the day in this regard, which is simply a mechanism that guarantees electricity purchases from an alternative source at a fixed rate that makes it competitive with fossil energy (see Jacobs 2010). This and other policy tools have contributed to a total installed solar PV capacity of about 69 gigawatts (GW) at the end of 2011 (EPIA 2012). This is noteworthy, considering it was 40 GW in 2010 and 23 GW in 2009. Europe dominates the PV market, with installed capacity at 51 GW (mostly from Italy and Germany), while Japan (5 GW), the United States (4.4 GW), and China (3.1 GW) represent the next largest markets.

This 2011 installed capacity is equivalent to about 69 commercial-scale (one GW) coal-fired or nuclear power plants. Unfortunately the installed capacity is a highly misleading number for solar (as well as wind, which we will talk about later). Installed capacity refers to how much electricity will be generated under ideal operating conditions. Solar, however, is an intermittent source. Daylight hours, of course, average only half the time, but even during daylight, the sun can be too low in the sky to produce significant power. Clouds disrupt insolation arriving at the solar facility. So, solar power only produces a fraction of the time. All in all, solar power can operate with load factors ranging from 10 to 25% of installed capacity. (Load factor is the amount of power actually produced on average, divided by the installed capacity.) In contrast, conventional coal and nuclear power plants operate with load factors approaching 90%. With this in mind, the 69 GW of installed capacity of solar photovoltaics actually produces the same amount of energy as about 12 coal-fired power plants, rather than 69, if we assume an average load factor of 17%. The growth in solar production, especially in the past four or five years is still encouraging, but needs to be viewed in the context of actual productive potential.

As we stated earlier, CSP technologies have been developed more extensively for large-scale applications. They are dependent on locations with consistent, intense sunlight, with desert areas typifying prime locations. According to the industry resource, CSP World, the end of 2012 saw a total installed capacity of about 2.5 GW, with most of that capacity in the United States and Spain (CSP World 2013). The smaller-scale applications, such as the one mentioned above developed by MIT, can be amenable to use in a broader range of climatic conditions. In many African communities, for instance, people are accustomed to generally unreliable power from the grid or to no electric power access at all, which is likely to make a somewhat intermittent source still acceptable. Recently, a prototype system has been installed in Lesotho, with five additional systems soon to be placed at clinics across that country (MIT 2006).

3.4 BIOMASS AND BIOFUEL RESOURCES

Most growing plant matter uses solar energy to combine carbon dioxide (CO_2) drawn from the atmosphere with water to store energy in the chemical bonds of organic molecules and to release oxygen. The organic molecules form the basis of biomass materials, which can be burned to release the energy stored. Fossil fuels are the remains of biomass long buried and altered by heat, pressure, and geochemical processes, which have merely been stored in the Earth for millions of years. Thus, biomass can be considered a very broad category, but the term is really only applied to biomass material grown during current lifetimes.

As such, biomass is popularly defined as "trees, grasses, agricultural crops, or other biological material" (NREL 2011). It is used in solid form for cooking and heating as firewood, charcoal, crop wastes, and dung, or converted into liquid or gas in order to also produce heat or to produce fuels, electricity, and other chemicals. They can be derived from a number of crop-based feedstocks, undergoing a number of conversion routes.

As is seen in Figure 3.6 there is a wide range of potential biomass feedstocks and routes that then can follow to end-use energy production. We, as do many authors, group them into three general categories. Lignocellulosic material includes wood and other solids. Crops and waste include sugar cane, corn and algae crops that yield sugars, starches or oily products, as

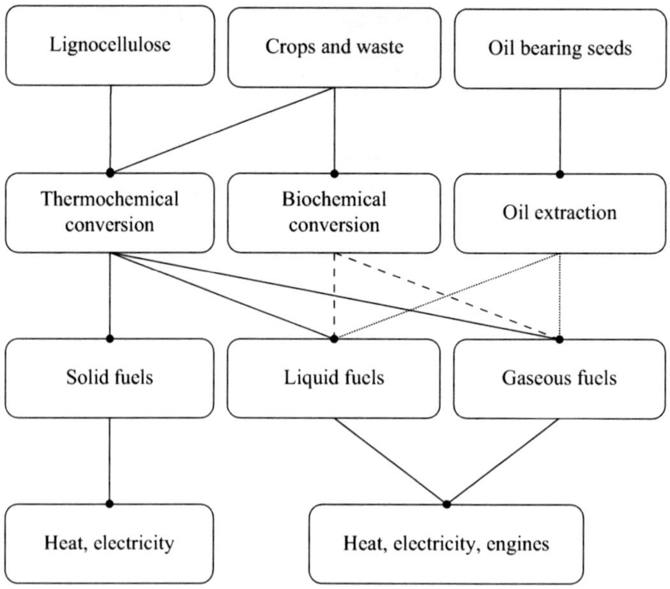

Figure 3.6. Biomass-to-fuel feedstock conversion routes.
Source: After FAO (1986) and UNDP (2000).

well as sewage, manure and municipal waste streams that are rich in organic content. Oil-rich seeds get their own category, in part due to having their own processing paths. Lignocellulose and crops and wastes can undergo basic chemical and thermal reactions to break down the complex organic molecules into simpler, easier burning compounds. Another route to break down complex molecules that is most pertinent to the crops and wastes is employing biological agents, such as yeast or anaerobic bacteria to effect the decomposition. For oil-rich seeds, the oils are simply extracted, such as by pressing. All of these conversion processes are meant to yield liquid, solid, or gaseous fuels that are suitable to burn to produce process heat, to generate electricity, or to run in engines.

Engineered biofuels liquids like fuel alcohol and biodiesel are gaining wider attention in the market, and are merely sourced from biomass that is grown specifically to be used as an energy crop. Currently, these biofuels represent a very small fraction of the biomass resource, but are forecasted to grow significantly for substituting fossil fuels and petrochemicals. Indeed, biofuel liquids have broad appeal for a number of reasons. First, as liquids, they can be put into culturally established modes of transportation. This is a key point that should not be overlooked—human resistance to change is a strong one and for the market to be able to provide a fuel that can be used very similarly to the current standard is extremely important to new fuel adoption. Second, biofuels are "dense enough" from an energy content basis to be realistically substitutable with respect to petroleum motor fuels. Third, which we contend to be extremely problematic, is that they are considered to be "green" because biofuels may burn cleaner than fossil fuels and because, theoretically, the CO_2 emitted by the combustion of these fuels can be recycled by new crops grown in the carbon cycle. Lastly, and most subjectively, it appears to liberate us from our oil "addiction." More questionable in terms of sustainability is the deliberate agricultural production of crops to use as fuels. It has received a great deal of attention and support, with further consideration provided below.

3.4.1 ETHANOL

Ethanol, which has gained wide attention in Brazil and the United States, is a product of fermentation of sugars by yeast. The yeast quickly becomes poisoned by the ethanol they produce, and typically die off when the alcohol concentration reaches 11% or 12%, with the remainder being mostly water. This means that the product must be distilled in an energy intensive process to separate the alcohol from the system. In Brazil, ethanol is most often derived from sugar cane, while in the United States it is most often derived from corn. These feedstocks should be distinguished.

Sugarcane contains sugar, which can be directly harvested and be processed into syrup, and then be easily converted to alcohol. Because the sugar is readily accessible, the process from "field to gas tank" makes sugarcane ethanol net energy positive (Macedo et al. 2008). Corn, on the other hand, consists primarily of starches, which must first be combined with amylase to break apart the starch polymers into separate sugar molecules (glucose) before being consumed by yeast to then make alcohol. Considering this and the energy required to grow the corn, transport the product, and distill fuel, several authors claim that the process has a net energy loss, while several others claim a net energy gain, due to byproducts that can be used to feed livestock (see Pimentel 2003; Shapouri et al. 2002). Balancing the biases of both advocates and detractors, it would appear that the net energy is approximately zero. Indeed, in a relatively balanced article, UC Berkeley researcher Kammen quantifies an error range with the estimation of net energy gain calculation that includes zero (Kammen 2007). In addition to this, based purely on land availability and other competing uses, ethanol production from this crop really should not exceed 10% of total gasoline at current rates of consumption. Thus, whether it has a slightly positive energy balance or slightly negative, corn ethanol appears to be at best a marginal energy source.

A stipulation to the problems with corn and other plant-based crops for fuel consumption (other than sugar cane) is the promise of using cellulosic materials in waste-for-fuel alcohol production. The appeal of this source (drawn from residual leaf, stalk, grass and tree material), especially in a nation like the United States, is that cellulosic materials could yield enough alcohol content to offset 30% of gasoline's market share (NREL 2007). But, as with corn fuel crops, cellulosic feedstocks require engineering—first to break down the fibers made of hemicellulose and cellulose into sugars and then to convert those sugars into alcohol. As of this writing, this engineering problem has yet to be solved on a practical basis. While a good deal of energy must be applied in the transformation of cellulosic material to ethanol and the subsequent refining to obtain pure ethanol, the energy inputs to grow the crop do not need to be counted against cellulosic ethanol, since the cellulosic material represents wastes from crops being grown for other purposes. Moreover, much of this cellulosic material is used for other energy and nonenergy applications, so it seems unlikely that all of the residual agricultural waste could be completely allocated to ethanol production.

3.4.2 BIODIESEL

Biodiesel, like ethanol, can be produced from crops, in this instance crops rich in vegetable oil, or from waste streams. Biodiesel can be blended directly with No. 2 Diesel if treated thermally or it can be treated chemically, through the process of transesterification, to make it useable directly (AFDC 2012). Used vegetable oil is a popular waste stream to use for this purpose.

As long as it is processed close to the point of production, it has the potential to be beneficial both environmentally and in terms of resource balance. The problem is that waste streams can typically only account for a small proportion of diesel demand. This is especially problematic in regions where gasoline, but not diesel, is more extensively utilized. There is also a potential problem that some waste for fuel may compete with food for animal livestock.

Additionally, if processed from soy crops, for example, the problem then relates to land use needs as it takes more land to produce equivalent bioenergy from soy compared to corn. Also, soy is a genetically engineered crop, and if we begin to select varieties that are heartier for conversion to fuels side by side with growing soy for food, there may be impacts on humans. The effects that genetically modified crops will ultimately have on the general population's health are unclear, so it would be wise to administer a certain degree of caution when advocating soy or any other heavily used food crop as a viable fuel crop.

Algae is another prospective feedstock crop for biodiesel production, which holds intriguing promise, because algae is abundant, extremely oil rich, and it can have extraordinarily high yields per acre, even being grown on trays in towers. CO_2 can be bubbled through the water on the trays to feed the algae. The main problem that exists with algae, though, is that concentrated CO_2 is needed to run the process efficiently. That can be an obstacle, but has the potential to be an opportunity as well. If algae facilities are co-located with major combustion facilities (such as a coal-fired power plant), the potential exists to tap the CO_2 emissions to feed the algae growth. This cannot realistically be viewed as sequestration, since the carbon will be emitted again when the algae biodiesel is burned, but it does double the energy use for each unit of CO_2. This has not been demonstrated in any commercial-scale applications as of this writing (see Scott et al. 2010 for a complete overview).

Finally, some photosynthesis processes produce small amounts of hydrogen ions, especially in low oxygen environments. Though, the production of hydrogen currently tends to be in very low concentrations. Nevertheless, some hopes are placed on the production of hydrogen through optimized biotic processes of this sort (Twidell and Weir 2006).

3.4.3 BIOGAS

One of the most sustainable uses of biomass would be to tap waste streams for fuel, especially biogas derived from sewage and municipal waste. Both of these sources inevitably produce methane as the organic waste material decays under anaerobic conditions. This process even occurs in the stomachs of ruminants, such as cattle. It happens naturally when any organic material is buried and is an early part of the formation of fossil fuels as well. The trick is simply to collect the methane, or biogas, and perhaps accelerate the decay to enhance methane production. This is a classic win-win situation, as the fuel is drawn from a waste stream, which would otherwise release methane to the atmosphere, which is a potent greenhouse gas.

Sewage systems generate methane naturally as do garbage dumps and manure tanks. Trapping the gas and burning it can be a very effective strategy, which also reduces the amount of greenhouse gas methane escaping to the atmosphere. Garbage dumps, primarily in the Developed World, are increasingly being fitted with pipes in the dump to collect the gas. The Mill Seat Land Fill, near Rochester, New York generates nearly 6 MW of electricity from landfill methane (NY DEC 2011). To achieve this level of production, they add sludge from the local sewage treatment plant, which is rich in the methanogenic microbes.

3.4.4 BIOMASS AND BIOFUELS DISTRIBUTION AND PRODUCTION

In terms of pure potential for the usage and cultivation of biomass for energy use, the world would rehquire a reasonable availability of arable land to grow and harvest feedstocks. Some regions will have more availability to arable lands than others, with a distribution looking something like that depicted in Figure 3.7.

According the International Energy Agency (IEA), biomass energy supplies about 50 Quads annually, representing about 10% of global energy consumption (IEA 2009). Indeed, a recent tally of biomass as a primary fuel reveals that forests and agricultural residues represent most of the biomass used in the world (Figure 3.8). This is mostly for heating and cooking.

Engineered biofuels represent a growing percentage of the biomass energy supply. From a production standpoint, global ethanol production as of 2011 was about 22.3 billion gallons (84.5 billion liters), while global biodiesel production was five billion gallons (GFRA 2012; Worldwatch Institute 2012). Finally, according to a recent Pike Research report (available at pikeresearch.com), biogas in 2012 was about 14 GW of installed power, both distributed and grid-tied. Of course, the sheer volumetric numbers do not tell the whole story, much like installed solar capacity does not tell the whole story once accounting for load factor. For ethanol in particular, the 22.6 billion gallon number must be reduced by a little over one-third to account for the energy density of the fuel with respect to gasoline. In that case, 22.6 billion reduces to around 15 billion gallons.

To gain perspective on what this means for future production and competing needs, let us look back briefly at the United States corn circumstance in 2010. The production of ethanol in the United States was estimated to be 13.23 billion gallons in 2010, which represented 9.5% by volume of gasoline consumption. However, since ethanol's energy density is only about

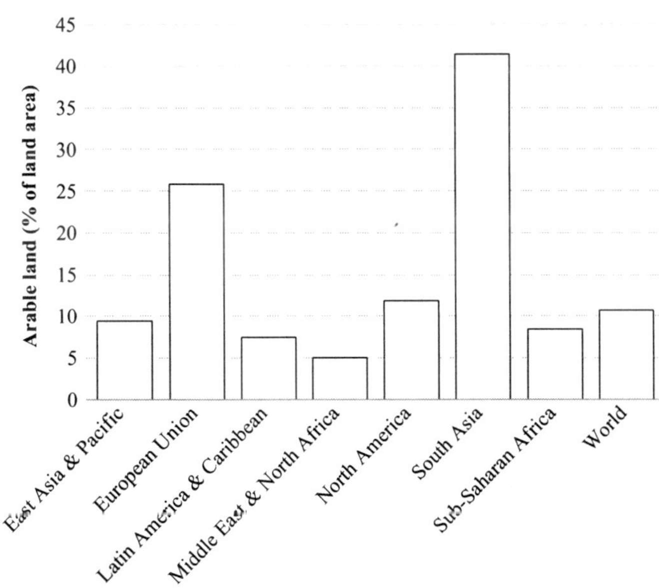

Figure 3.7. Global availability of arable land, end 2008.
Source: World Bank (2008).

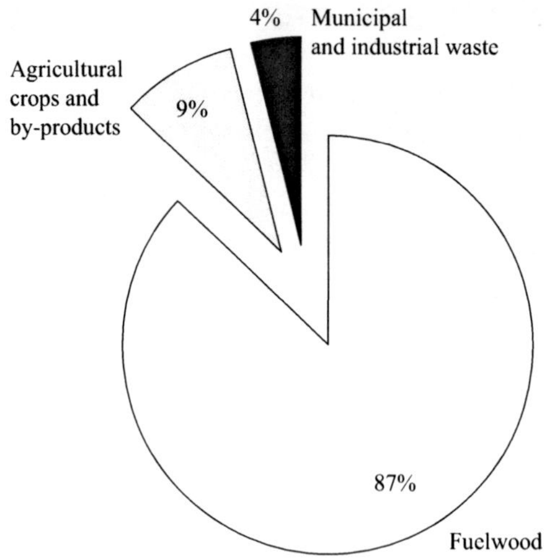

Figure 3.8. Global biomass energy supply by percentage of source.
Source: IEA (2009).

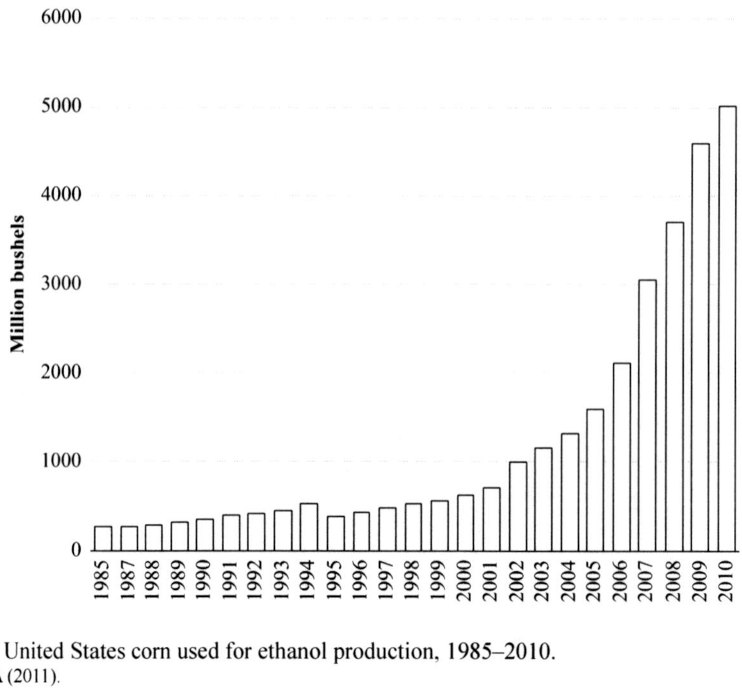

Figure 3.9. United States corn used for ethanol production, 1985–2010.
Source: NCGA (2011).

60% of gasoline's (in terms of energy content per gallon), the energy delivered by ethanol was actually 5.7% of the energy delivered by gasoline. In 2010, about 38% of America's corn crop was used to produce 5.7% of America's automotive fuel by energy content (see Figure 3.9). Under existing technologies, it would require 12 times as much land to produce as is currently used for all of America's corn production. One might argue that the yields for corn production have been increasing, so the land use would not have to increase by that much. Thus, it might not take 12 times more land than we use for all of corn production, but it is unreasonable to assume a 1,200% increase in corn yields (droughts tend to put a damper on that, among other uncontrollable events). Corn ethanol will have to face issues of competition for cropland. This is highly problematic as a sustainable energy source. The competition became more apparent in the drought of 2012.

3.5 HYDROPOWER

Hydropower is currently the largest producer of the renewable energy sources. It utilizes gravitational, mechanical, and kinetic energy to produce power. Recall that the ultimate source of hydropower is the sun. As solar energy heats Earth's surface, it evaporates water, which rises in the atmosphere to form clouds. As the clouds are carried over high ground, cool temperatures cause the vapor to condense. The resulting rainfall or snow accumulates and ultimately flows into rivers that in turn flow toward the sea. This is known as the water cycle (USGS 2012).

Water power has been utilized for centuries, going back at least 2000 years to the first known water wheel, the Noria. It was developed largely to make use of the energy of a flowing stream to lift water to a higher elevation. Pails were connected to the outside of the wheel. As the water flowed into one pail, it turned the wheel, bringing the next pail down to be filled, while lifting the full pails. Once this simple technology came into use, innovations marched on, leading to geared systems as represented by the early Vitruvian waterwheel. The gears provided the capacity to change the axis and speeds of rotation. The gearing enabled a wide range of new technologies to build on the energy provided by hydropower. Milling grain was greatly facilitated by water power. In spite of reports that millers resisted the introduction of water mills because they did not want to move their operations to streams and rivers, effectiveness of the new energy source carried the day, and by the sixth century AD water mills began to appear along many rivers and stream in Europe (Cook 1976). Eventually towns grew up alongside streams, in part to take advantage of the available hydropower to mill grain and form other mechanical operations.

The evolution of hydroelectric power plants from water mills was fairly straightforward. The same rotational motion of the wheel that turned a millstone now provides the basis for generating electricity. As seen in Figure 3.10, water is typically directed through the channel (sluice) or a pipe (penstock) to ensure the water's force is delivered to the wheel, rather than diffusing around it. The construction of the dam confines a portion of the moving water and provides a large head (height from which the water falls). The head determines the pressure or force per unit area exerted by the water at the bottom of this fall across the blades of the device. Even in a rapidly flowing stream, the pressure across a turbine is caused merely by the height difference from immediately above the device to immediately below it. A dam concentrates the height change of a significant length of the river to a single point.

Small-scale hydropower projects may also employ a small dam, but are often "run of the river." These projects are receiving some favorable attention in regard to reduced environmental impact and highlighting the advantages of decentralized energy. This can include micro hydropower at a scale of less than 100 kW per installation and mini hydro at scales up to 1 MW (Paish 2002). The disadvantage to small-scale hydropower is that it tends to lack the economies of scale associated with large projects, where all of the equipment and expenditures are concentrated at a single site and the large project is able to enjoy the benefits of pricing for large purchases.

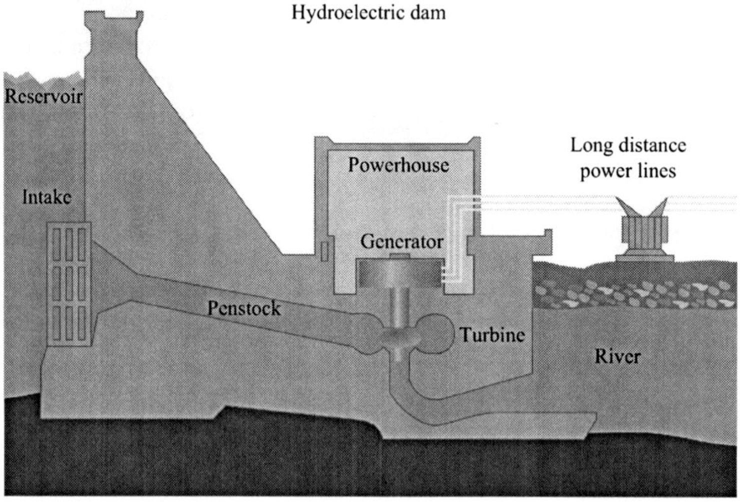

Figure 3.10. Hydroelectric dam.
Source: Tomia/CC-BY-2.5, Wikimedia Commons.

3.5.1 HYDRO POTENTIAL DISTRIBUTION

Since water movement is a surface phenomenon, exploration is not a problem for hydropower. What is an issue is the engineering challenge of directing surface flow into areas that can utilize standard hydropower principles. Many of the best sites for large-scale hydropower have already been used in the industrialized world, limiting the opportunity to expand this resource substantially. Where the opportunity does exist, the issue of ecosystem disruption has begun to raise serious environmental questions and may well limit the extent of future large-scale hydropower development. Small-scale and run-of-the-river applications have extensive availability for distributed, decentralized energy, perhaps especially in the Developing World. At least from a technical stand, there are still many opportunities to exploit the water-to-power resource, as shown in Figure 3.11.

3.5.1.1 Is Hydropower a Renewable Resource?

Some authors and government agencies now exclude, or least purposely separate large-scale hydropower from the list of renewable energy sources. This would seem yet another indictment of the use of the "renewable" terminology. It would seem that the only real difference between large-scale hydropower and the other renewable resources is that it is coming under increasing scrutiny for environmental and social impacts. These are legitimate sustainability issues, but to say that they affect whether something is renewable or not seems clearly to show how this resource/abundance issue has been conflated with unrelated issues.

3.5.2 TIDAL AND WAVE POWER

While most hydropower resources are based on both solar power and gravity, tidal power can be viewed as strictly a gravity-based energy resource. It taps the kinetic energy of water that is

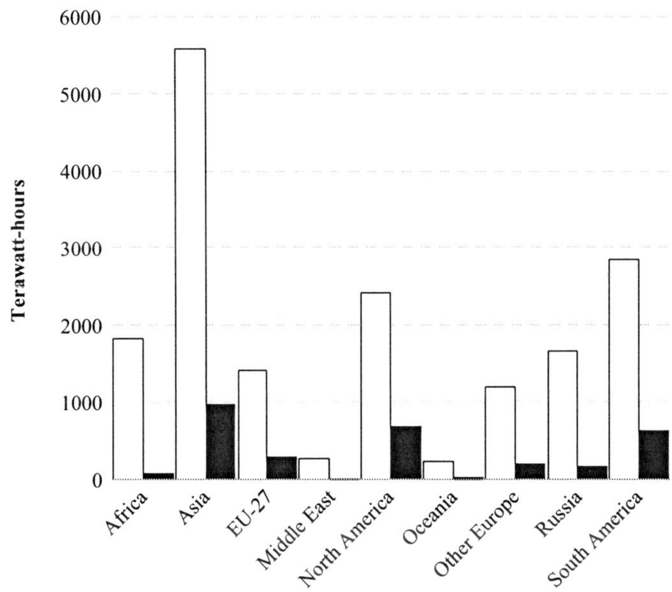

Figure 3.11. Hydropower: Annual technically exploitable capability (white) and actual generation (black) by region, end 2008.
Source: WEC (2010).

raised above its base level by the gravitational attraction of the moon. Tides are actually a bulge of water drawn toward the moon. As Earth rotates, the bulge moves around the world. Depending on the shape of the coastal surface, some tides are more dramatic than others. Especially in locations with exaggerated tidal flux, it is possible to impound some of the water at high tide and channel it through a hydropower plant as the water falls back to the low tide level. This would be the barrage or dam approach, but these sites are quite limited (Baker 1991). Another technology is referred to as tidal fences, which span channels that experience fast moving tidal currents. Devices described as "looking like turnstiles" spin as the water flows by and generate electricity (Khalid et al. 2012). Finally, other tidal turbines may resemble wind turbines (described in the next section), with blades spinning on a horizontal axis. It should be noted that, because water is so much denser than air, even though the currents are barely moving at the minimum speed to start a wind turbine, tidal devices can capture as much energy as a wind turbine four times as large. Only a few prototype tidal power installations are operational as of this writing. The combined potential of ocean tidal and tidal current energy is estimated to be approximately 7,800 terawatt-hours (TWh) per year—that equates to about 0.9 TW, well under 10% of the global total energy demand, but still significant, especially in comparison to other renewable resources (OES 2011).

Wave power is another form of energy from the ocean, which is ultimately derived from solar power, in that wave action is largely caused by wind, which is, in turn, a result of the uneven heating of Earth's surface by the sun. It represents quite a large overall energy flux. Much of it, though, is very diffuse. Regions with significant, relatively continuous wave action still represent a potential sufficient to have drawn interest for several decades. The technical challenge is to transform the rise and fall of ocean currents into useful energy: primarily turning a turbine to generate electricity. A wide array of technologies is being considered, including oscillating water columns, oscillating bodies, and overtopping (OES 2011). Naturally, wave power is restricted to coastal regions, but it is possible to serve many people, as population centers also tend to concentrate along coasts.

3.6 WIND POWER

Wind power is very much like hydropower in that it harvests energy from flowing fluids—air, rather than water. This fluid flow is set in motion largely by solar power. The uneven heating of Earth's surface warms the air immediately above to varying degrees, driving air circulation across the planet. This uneven heating, coupled to Earth's Coriolis forces, sets much of the global wind patterns at altitudes above about 100 meters (328 feet). Below this altitude these wind patterns are slowed down and their directions further altered by Earth's rotation and varying surface features like mountains, hills, trees, and buildings. Localized wind action, concentrating the power, such as sea breezes and mountain winds can also contribute to wind patterns, especially when the larger scale wind patterns are lighter than normal (see Boyle 2004).

If you can imagine a sheet of air of known density with a known area traveling past an object at a certain velocity that you can measure, you can derive a relatively straightforward relationship for the wind power traveling past that object. Essentially, this relationship can be expressed as:

$$\text{Wind Power} = \frac{1}{2}(\text{air density}) \times (\text{size of wind-blown area}) \times (\text{air velocity})^3$$

In other words, wind power is proportional to:

1. The density of air
2. The area through which the wind is passing
3. The cube of the wind velocity

If the object experiencing that passing wind is configured in the right way, it can convert the moving kinetic energy into useful work. The very simple relationship above tells us that the available power from the wind at any location is dependent primarily on the wind velocity, then to a lesser extent the size of the windblown area (i.e., the total amount of wind hitting a surface), then to an even lesser extent the density of the air. Based on these principles, but before this relationship was ever derived, the wind resource was utilized for several centuries. It was first used for mechanical work to grind, mill, and pump water. Pumping water remained a long-standing use of wind power well into the 20th century—wind powered water pumps proliferated on American farms and ranches. (These wind pumps were being rapidly replaced with the convenience of electric submersible pumps even in the 1970s energy crisis era.)

Wind has also been used for millennia as an energy source for transport via sailing vessels. Indeed the clipper ships of the 19th century produced more power than even most large wind turbines today. Some authors suggest that, with modern materials and computer controls, even more effective sailing vessels can be built. International shipping could make a rather small dent in global energy consumption, but it may be useful, and probably would have a somewhat larger mitigation of environmental impact, as National Academies studies show a fairly significant amount of marine oil pollution resulting from normal tanker operations worldwide (United States regulations enacted in 1990 have almost eliminated what was a significant source here) (NAP 2003). Early wind power technologies employed very large sails that captured a great deal of the passing air. Those sails could be the canvas on sailing ships or the large arms of the picturesque windmills that Don Quixote mistook for giants.

3.6.1 WIND TURBINES

In more recent times, wind has been transformed into a source to generate electricity at an industrial scale. Now the blades tend to be long and slender to operate in the very high wind speeds encountered at higher altitudes. Wind experienced the greatest rate of growth at the turn of the 21st century of any of the renewable, flow-limited energy sources. Of course, a large growth rate for an extraordinarily small resource still equates to small production in the global context, but many believe the world is just getting started with the development and deployment of wind turbine technologies. Generally, wind turbines are classified by whether they rotate on a horizontal or vertical axis and by their output (EWEA 2009). Horizontal axis wind turbines are the dominant technology for utility-scale production and their rated output is typically classified by the length of their turbine blades or equivalent swept area (see Figure 3.12).

The late 20th century saw ever-increasing scales of wind turbines, with some in place and producing 5 MW of power, and each blade arm more than 60 meters long, with even larger blade lengths on the horizon. And, while these are very impressive structures, it would take 200 of these huge 5 MW wind generators to match one relatively large coal or nuclear power plant. Ideal sites for such large-scale wind farms are relatively limited and often remote from population centers.

Also, in the beginning of the 21st century, there is some momentum in the opposite direction, with innovations being made at small-scale. Some of these are shrouded to increase the

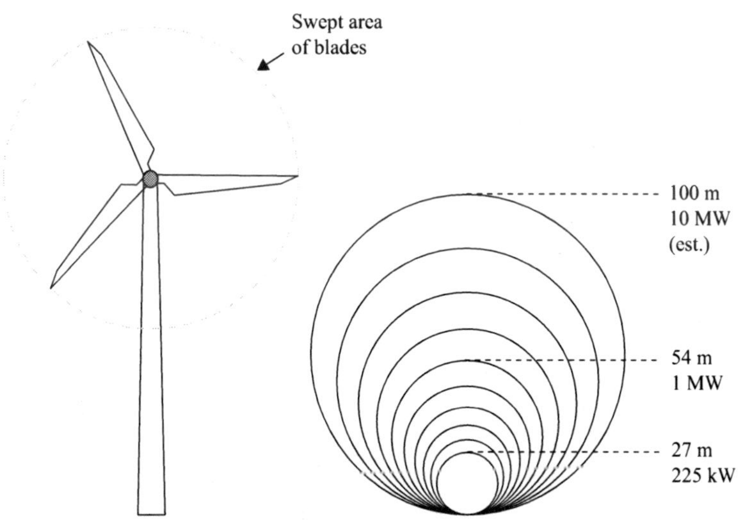

Figure 3.12. Depiction of a horizontal axis wind turbine (left, also showing the swept area of the turbine blades) alongside a turbine capacity chart (right, a visual depiction of blade/swept areas to maximum power output of a wind power system).

pressure drop across the blades. Others have introduced cowling to "trick" the wind to speed up. Some of these are actually roof-mountable, which makes them applicable anywhere that there is some wind. But this class of turbine still needs to address a number of issues, ranging from problems with vibrations (a huge problem when placing them on structures), to reliable power output at the very low cut-in wind speeds (the wind speed at which the turbine starts producing power). And, since the area of a small turbine will be small, it is still essential that the turbines are placed in areas with good wind, necessitating really good anemometry (the study of wind speeds) at the installation site. In addition to the technical obstacles that need to be addressed, the dissemination of this technology will begin to depend on a shift toward decentralized power.

3.6.2 *WIND DISTRIBUTION AND INSTALLED CAPACITY*

As with water, we know that wind is ubiquitous. But unlike hydropower, it is impossible to store the wind behind a dam and use it when needed. Thus, it is important to start off with a good estimation of wind speeds at the heights that the turbine blades will be installed. So, many efforts have gone forward to map the wind resource. Indeed the trend has been to map out the distribution of the wind resource at heights approaching 100 meters because the only technical factors that come into play in planning a project, then, are tower heights and blade lengths. As such, there have been a number of independent researchers and government agencies that have been doing this work (see Archer and Jacobson 2005). To get good local information on winds speeds, it remains essential to do a wind audit at the site of the turbine installation, at the exact height that the turbine blade will be located, since the performance of the blade is so sensitive to the wind speed (using the wind power equation above, you can see that the difference between 9 meters per second versus 10 meters per second could impact the expected power output by as much as 30%).

Finally, as with the solar resource, what we seem to want to know most about the wind resource is what the currently installed capacity is. According to the Global Wind Energy Council (GWEC), installed wind capacity at the end of 2012 was 282 GW (GWEC 2013). This is remarkable, because as recently as 2007, installed capacity was just less than 95 GW. Installed wind capacity really has grown quite substantially in a relatively short time, as depicted in Figure 3.13, where in 1996 cumulative installed capacity was recorded to be 6.1 GW compared to the 282 GW of 2012.

We, again, should be mindful of the load factor issue attached to installed capacity. Load factors for wind vary depending on location and whether the installation is onshore or offshore. A safe, perhaps even optimistic number for a load factor is about 0.30 for onshore installations, which represents the bulk of the installed capacity at the top of the 2012 list in China, the United States, and Germany. That means 282 GW of installed capacity is closer to 84.5 GW of practical production, which still is very good but is also pretty small compared to the roughly 2.6 TW (1 TW = 1000 GW) available from cumulative installed conventional thermal electricity operating at a load factor of 0.85.

3.7 GEOTHERMAL

The word geothermal literally means "Earth Heat." Technically, it is the only nondepletable resource that is not influenced by the sun (tidal energy is an artifact of the gravitational

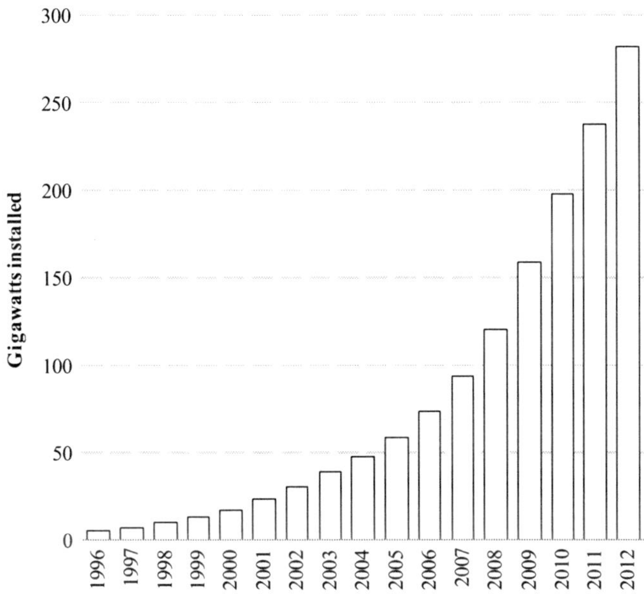

Figure 3.13. Global cumulative installed wind capacity, 1996–2012.
Source: GWEC (2013).

relationship with the moon, but ultimately that is affected by how Earth and its moon are gravitationally related to the sun). The heat of Earth's interior is substantially a product of the nuclear decay of naturally occurring radioactive elements and isotopes, while the crust is now understood to be a dynamic system, with massive tectonic plates floating on the molten mantle. In spots where the crust is thin or where the tectonic plates are crashing together, or pulling apart, some of the hot, molten rocks rise toward the surface.

A volcano probably is the most dramatic expression of heat escaping to the surface. If we could realistically tap into the energy of a volcano, we would have a tiger by the tail—an enormous energy source, but one that is neither uniform nor predictable. Hot springs, geysers, and fumaroles are three other dramatic, but less violent and less erratic examples of Earth's internal heat escaping to the surface. These three occurrences represent forms of geothermal energy that we do harness. Finally, ocean thermal may be considered a subset of geothermal, because it too takes advantage of natural temperature gradients near the surface of the Earth. In this case, though, the ocean depths provide the cold temperature to condense a highly volatile working fluid, while the warmer surface temperatures provide the heat source. So, it is important to note that energy is produced based on heat or temperature differences. Surface temperatures are hot compared to the ocean floor, but cold compared to steam coming from the boilers of a thermal power plant.

A geothermal resource can be described as a geologic reservoir from which the heat can economically be extracted (Rezaie and Aghajani 2013). This heat, depending on its characteristics, can then be utilized for thermal and/or electric power applications. A geothermal reservoir can also contain heat in the fluids that fill the pore spaces of the source rock in addition to the heat contained within the rock itself. The fluids that are used for producing electricity are available at depths of about a mile or more, whose energy is utilized by three basic power plants: (1) dry steam; (2) flash steam; and (3) binary cycle (GTO 2012). The choice of plant type depends on the fluid phase of the reservoir. A dry steam plant uses steam directly to turn a turbine.

A flash steam plant draws high-pressure hot water up to the surface and runs it through low pressure tanks to flash vaporize the water and then run it through turbines. (The cooling water is reinjected into the reservoir.) A binary cycle plant is similar to a flash steam plant; however, the hot water from the reservoir passes by a secondary fluid with a lower boiling point temperature. This secondary fluid vaporizes and is then run through turbines. Many potential geothermal areas contain moderately hot water (rather than "dry steam") and most future geothermal energy plants will likely employ binary cycle technology to generate electricity.

Although acquired through drilling very similar to oil and gas, the geothermal resource can be exploited much more sustainably than fossil fuels, because heat is constantly being generated in Earth's interior by decay of radioisotopes. As steam is produced to generate electricity, condensed water is reinjected to absorb more heat from the reservoir and delivered to the surface to generate more electricity.

3.7.1 GEOTHERMAL DISTRIBUTION AND INSTALLED CAPACITY

Since Earth's interior is known to be extremely hot, geothermal energy is ubiquitous. Thus, if we drill down into the surface anywhere on the planet, we will encounter increasing temperatures. Despite the ubiquity of the resource, though, most of the energy available that is actually accessible near the surface is very diffuse, except for the geothermal resources that are found near plate boundaries in an area that encircles the Pacific Ocean called the "Ring of Fire" (see Figure 3.14). This area is where most of the reserves are also located. (Though, there is hope that enhanced drilling technologies will expand where geothermal energy can be exploited.) It is in the Ring of Fire that we find the highest concentrations of hot springs, geysers, and fumaroles, where we can recover the extracted heat in the form of steam or hot water and use

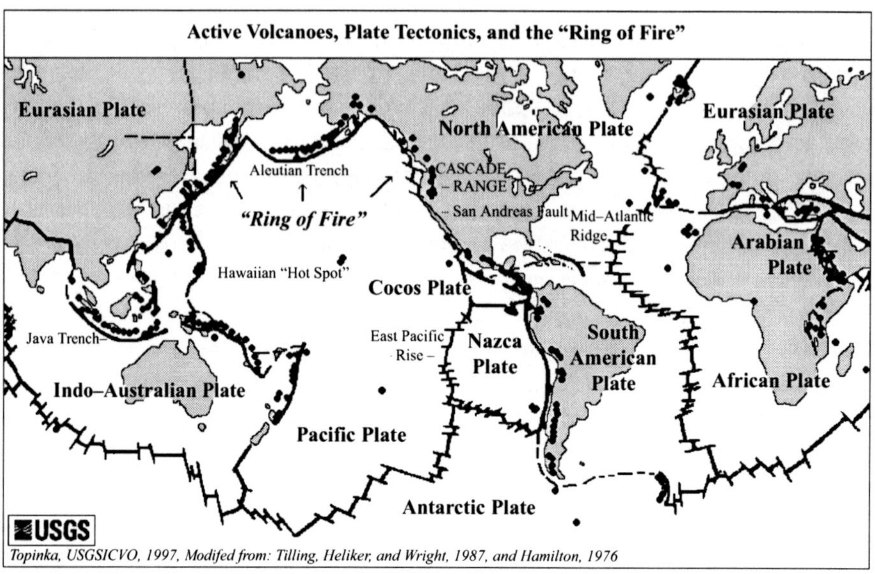

Figure 3.14. The "Ring of Fire," an area encircling the Pacific Ocean where most of the geothermal reserves are located.
Source: USGS (1997).

it to generate electricity and to heat structures. According to Stefansson (2005), the technical potential of this region is roughly 6.5 terawatts of power per year (thermal basis).

One of the largest opportunities for growth in geothermal electric power generation is in developing the technologies to exploit hot dry rock. Most commercial geothermal power projects exploit natural steam reservoirs, in which water exits in porous or naturally fractured rock that has been heated by unusually shallow occurrences of high temperatures. There is far more hot rock that does not contain steam in naturally occurring porous structures, than the special applications that are currently being tapped. Perhaps the advances in precisely controlled directional drilling and extensive hydraulic fracturing may offer opportunities to create extensive conduits in the abundant hot rocks to transport water and convert it to steam.

In terms of electricity production, the current installed capacity of geothermal globally is about 11 GW, with a distribution of 2010 installed capacity depicted in Figure 3.15.

3.7.2 DIRECT USE APPLICATIONS

In addition to geothermal usage for electric power, there exists broad potential for direct access in reservoirs of low to moderate temperatures (20°C–150°C). Hot water near the surface can be piped directly into buildings for a number of heating applications, including for district heating of municipal and residential buildings, for melting of snow on roadways, for growing food in greenhouses, and for a number of other agricultural needs (GTO 2013).

Also, in shallow ground (roughly 10 ft) there are opportunities to use the geothermal resource. Almost everywhere at this depth, geothermal temperatures are nearly constant, ranging from 10°C to 16°C. Using a heat pump, the temperature differences between the surface and

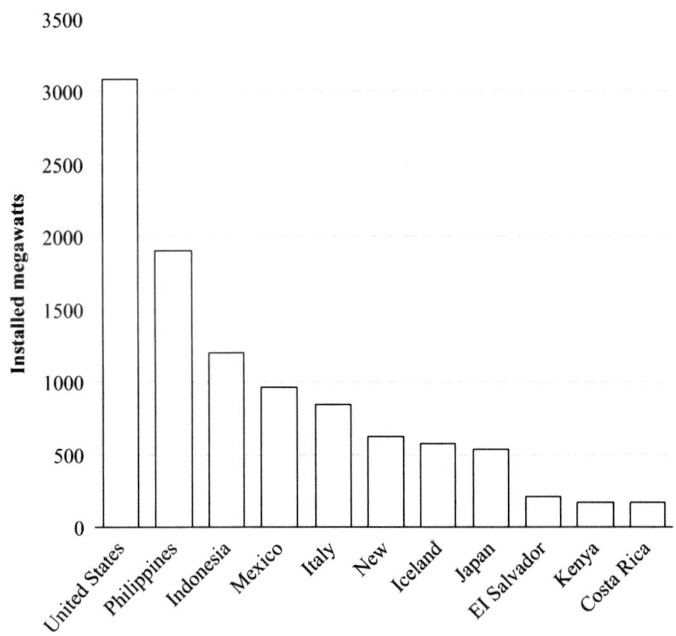

Figure 3.15. Distribution installed geothermal capacity end 2010.
Source: GEA (2012).

this shallow depth are sufficient to aid in home heating and cooling applications. In the winter, heat from the relatively warmer ground can be pumped up and through a heat exchanger to help with heating needs. In the summer, heat from the home can be pulled down to the relatively cooler ground to help cool the house. This has very large potential, especially in regions where surface-to-ground temperature differences are always large.

REFERENCES

AFDC (Department of Energy Alternative Fuels Data Center). 2012. "Biodiesel Production and Distribution." http://www.afdc.energy.gov/fuels/biodiesel.html

Archer, C. L., and M. Z. Jacobson. 2005. Evaluation of global wind power. *Journal of Geophysical Research, 110* (D12), D12110. doi: http://dx.doi.org/10.1029/2004JD005462

Baker, C. 1991. Tidal power. *Energy Policy. 19* (8): 792–797. doi: http://dx.doi.org/10.1016/0301-4215(91)90049-T

Boyle, G. B. ed. 2004. *Renewable Energy*. New York: Oxford University Press.

Chiras, D. 2009. *Power from the Sun*. Gabriola Island: New Society Publishers.

Cook, E. 1976. *Man, Energy, Society*, San Francisco: W. H. Freeman.

CSP World. 2013. "CSP 2012, a review." http://www.csp-world.com/resources/csp-2012-review

DOE (Department of Energy). 2011. Energy 101: Daylighting. http://youtu.be/-7EG4d-W4W8

EERE (Department of Energy Energy Efficiency & Renewable Energy). 2012. "Linear Concentrator Systems for Concentrating Solar Power." http://www.eere.energy.gov/basics/renewable_energy/linear_concentrator.html

EPIA (European Photovoltaic Industry Association). 2012. "Global Market Outlook for Photovoltaics until 2016." http://www.epia.org/news/publications/global-market-outlook-for-photovoltaics-until-2016/

EWEA (European Wind Energy Association). 2009. *Wind Energy—The Facts: A Guide to the Technology, Economics and Future of Wind Power*. London: Routledge.

FAO (Food and Agriculture Organization of the United Nations). 1986. "Power for Pumping." In *FAO Irrigation and Drainage Paper 43: Water lifting*. http://www.fao.org/docrep/010/ah810e/AH810E13.htm

GEA (Geothermal Energy Association). 2012. "Geothermal: International Market Overview Report." http://www.geo-energy.org/reports.aspx

GFRA (Global Renewable Fuels Alliance). 2012. "Global Ethanol Production to Reach 85.2 Billion Litres in 2012." http://www.globalrfa.com/pr_062612.php

GTO (Department of Energy Geothermal Technologies Office). 2012. "Energy Basics: Electricity Generation." http://www1.eere.energy.gov/geothermal/powerplants.html

GTO (Department of Energy Geothermal Technologies Office). 2013. "Direct Use of Geothermal Energy." http://www1.eere.energy.gov/geothermal/directuse.html

GWEC (Global Wind Energy Council). 2013. "Global Wind Statistics 2012." http://www.gwec.net/publications/global-wind-report-2/

IEA (International Energy Agency). 2009. "Bioenergy—A Sustainable and Reliable Energy Source." IEA Bioenergy. www.ieabioenergy.com/DownLoad.aspx?DocId=6494

Jacobs, D. 2010. "Introduction to feed-in tariffs." At renewableenergyfocus.com. http://www.renewableenergyfocus.com/view/13313/introduction-to-feed-in-tariffs/

Kammen, D. M., A. E. Farrell, R. J. Plevi, A. D. Jones, G. F. Nemet, and M. A. Delucchi. 2007. "Energy and Greenhouse Impacts of Biofuels: A Framework for Analysis." Research Round Table on Biofuels: Linking Support to Performance. Paris. http://pubs.its.ucdavis.edu/download_pdf.php?id=1131

Khalid, S. S., Z. Liang, and N. Shah. 2012. Harnessing tidal energy using vertical axis tidal turbine. *Research Journal of Applied Sciences, Engineering and Technology. 5* (1): 239–252.

Kiehl, J. T., and K. E. Trenberth. 1997. Earth's annual global mean energy budget. *Bulletin of the American Meteorlogical Society*. 78: 197–208. doi: http://dx.doi.org/10.1175/1520-0477(1997)0783C0197:EAGMEB3E2.0.CO;2

Macedo, I. C., J. E. Seabra, and J. E. Silva. 2008. Green house gases emissions in the production and use of ethanol from sugarcane in Brazil: The 2005/2006 averages and a prediction for 2020. *Biomass and Bioenergy*. *32* (7): 582–595. doi: http://dx.doi.org/10.1016/j.biombioe.2007.12.006

MIT (Massachusetts Institute of Technology). 2006. MIT News, "Students win grant for Lesotho solar generator." http://web.mit.edu/newsoffice/2006/lesotho.html

NAP (National Academies Press). 2003, "Oil in the Sea III: Inputs, Fates and Effects." http://www.nap.edu/openbook.php?isbn=0309084385

NASA (National Aeronautics and Space Administration). 2012. "Sun: Facts & Figures." http://solarsystem.nasa.gov/planets/profile.cfm?Object=Sun&Display=Facts&System=Metric

NCGA (National Corn Growers Association). 2011. "World of Corn 2011." http://www.ncga.com/worldofcorn

NREL 2001. "Passive Solar Design for the Home." DOE/GO-102001-1105. http://www.nrel.gov/docs/fy01osti/27954.pdf

NREL. 2003. "Heat Your Water with the Sun." DOE/GO-102003-1824. http://www.nrel.gov/docs/fy04osti/34279.pdf

NREL. 2007. "Research Advances: Cellulosic Ethanol." NREL/BR-510-40742 http://www.nrel.gov/biomass/pdfs/40742.pdf

NREL. 2011. "Glossary of Biomass Terms." http://www.nrel.gov/biomass/glossary.html

NY DEC (New York State Department of Environmental Conservation). 2011. "Landfill Gas Recovery Data." http://www.dec.ny.gov/chemical/48873.html Note that many such landfills are shown to generate electricity from landfill gas, with some even larger producers than Mill Seat.

OES (International Energy Agency Ocean Energy Systems). 2011. "Waves" and "Tidal and Currents." http://www.ocean-energy-systems.org/ocean_energy

Paish, O. 2002. Micro-hydropower: status and prospects. *Proceedings of the Institution of Mechanical Engineers, Part A: Journal of Power and Energy*. *216* (1): 31–40. doi: http://dx.doi.org/10.1243/095765002760024827

Pimentel, D. 2003. Ethanol fuels: energy balance, economics, and environmental impacts are negative. *Natural Resources Research*. *12* (2): 127–134. doi: http://dx.doi.org/10.1023/A:1024214812527

Rezaie, M., and H. A. Aghajani. 2013. A new combinational terminology for geothermal systems. *International Journal of Geosciences*. 4: 43–48. doi: http://dx.doi.org/10.4236/ijg.2013.41005

Scott, S. A., M. P. Davey, J. S. Dennis, I. Horst, C. J. Howe, D. J. Lea-Smith, and A. G. Smith. 2010. Biodiesel from algae: challenges and prospects. *Current Opinion in Biotechnology*. *21* (3): 277–286. doi: http://dx.doi.org/10.1016/j.copbio.2010.03.005

Shapouri, H., J. A. Duffield, and M. Q. Wang. 2002. "The Energy Balance of Corn Ethanol: An Update." US Department of Agriculture Agricultural Economic Report No. 814. http://purl.umn.edu/34075

Smil, V. 2008. *Energy: A Beginner's Guide*. Oxford: Oneworld Publications.

St. John, J. 2009. GreentechSolar. "Lockheed Martin, Starwood to Build 290MW Solar Thermal Plant in Arizona." http://www.greentechmedia.com/articles/read/lockheed-martin-starwood-to-build-290mw-solar-thermal-plant-in-arizona/

Stefansson, V. 2005. World geothermal assessment. *Proceedings World Geothermal Congress*. Antalya, Turkey, 24–29. April, 2005.

Twidell, J., and T. Weir. 2006. *Renewable Energy Resources*. New York: Taylor and Francis.

UNDP (United Nations Development Programme). 2000. "World Energy Assessment: Energy and the Challenge of Sustainability." http://www.undp.org/content/undp/en/home/librarypage/environment-energy/sustainable_energy/world_energy_assessmentenergyandthechallengeofsustainability/.

USGS (United States Geological Survey). 1997. "Active Volcanoes and Plate Tectonics, 'Hot Spots' and the 'Ring of Fire.'" http://vulcan.wr.usgs.gov/Glossary/PlateTectonics/Maps/map_plate_tectonics_world.html

USGS (United States Geological Survey). 2012. "The Water Cycle." http://ga.water.usgs.gov/edu/watercycle.html

WEC (World Energy Council). 2010. "2010 Survey of Energy Resources." http://www.worldenergy.org/documents/ser_2010_report_1.pdf

World Bank. 2008. "Data: Agricultural land (% of land area)." http://data.worldbank.org/indicator/AG.LND.AGRI.ZS

World Watch Institute. 2012. "Biofuels Regain Momentum." Vital Signs. http://vitalsigns.worldwatch.org/vs-trend/biofuels-regain-momentum

CHAPTER 4

Energy Consumption in Economic Sectors

This chapter describes how energy is processed and consumed in the primary sectors of industrial economies.

4.1 BROADLY CHARACTERIZING ENERGY CONSUMPTION

Consuming energy, in the strictest physical sense, is a misnomer. What is consumed, of course, is an energy resource like coal or oil, commonly referred to as "primary energy." The energy stored within these resources can be exploited—through some energy conversion process—to extract more useful heat, light, and/or kinetic energy. This useful energy then is used to provide both essential and nonessential services to a society. So, when we talk about energy consumption, we are really talking about our utilization of fuels that can be ultimately converted into the many kinds of services they provide. It is not the energy itself that we seek to consume.

Also, the form of energy is significant. Electricity, which can be derived from fossil and nonfossil energy, is extremely versatile, while gasoline and diesel fuels have very high energy and power densities, which are essential features for storage, such as is required in vehicular applications. Natural gas is extremely clean and burns very efficiently for heating and cooking. Conversely, solid fuels tend to be less clean, less efficient, and more difficult to deliver to individual consumers. This helps to explain the efforts to produce synthetic crude oil or synthetic gas from coal, or even ethanol from crops. Liquid fuels offer greater utility.

Furthermore, consumer needs dictate what utility is most important. For example, it is possible, but not very efficient, to light a room by burning fossil fuels. Likewise, it is not feasible to run a TV or computer directly on a combustion fuel. Additionally, liquid combustion fuels offer significant utility in vehicular applications. So while consumption can be defined as "the using up of a good or a service having an exchangeable value," when contemplating energy consumption, the degree to which that value can be quantified or "monetized" depends very much on the consumer's desire for certain goods and services balanced by their desire for equity and environmental protection as a result of consuming energy resources to get that good or service. That ultimately dictates the amount and types of primary energy to be used by society, and how they can be utilized within the consumption sectors of an economy.

4.2 ENERGY CONSUMPTION IN INDUSTRIALIZED SOCIETY

In Industrial and Post-Industrial Nations like those in North America and in Europe, or in Industrializing Nations like China, India, and Brazil, there are universally recognizable economic sectors that consume energy for productive purposes. In the United States, these sectors are organized into four primary ones and are described by the Energy Information Agency (EIA) as follows:

- The electric power sector, in which a great deal of energy is transformed from primary sources to the highly efficacious energy carrier, electricity. Many authors and reporting agencies do not always treat this as a separate sector, since it does not represent an end use of energy. That is, the electricity produced in this sector is in turn consumed in each of the other sectors. Nevertheless, the transformation to electricity is sufficiently important and represents sufficient energy use that we choose to treat it separately.
- The transportation sector, which includes all forms of land, sea, and air transportation, both for people and for goods. It is by far the largest consumer of petroleum, which is the largest energy provider in the world today.
- The industrial sector, which refers to the energy used in making all kinds of goods, including through manufacturing, agriculture, mining, and construction activities.
- The residential and commercial sectors, which include the energy used to provide climate control, light, cooking, and hot water in all kinds of buildings. Some countries separate these sectors, but we consider them together, as the technologies employed have tremendous overlap.

These sectors are described in further detail below.

4.3 THE ELECTRIC POWER SECTOR

Electricity is essential to powering industrialized and industrializing economies. It affects nearly every aspect of daily modern living, and indeed is an essential element to the industrial, residential, and commercial sectors of society. It is almost unimaginable to lack access to electricity for heating, cooling, lighting, working on computers, and watching TV shows. As Smil (2008) puts it, electricity is the "most convenient, most flexible, and most useful of energies." Flip a light switch and a whole room can be illuminated. Plug in an air conditioner, turn it on, and a hot, sweltering day becomes manageable for working, playing or even sleeping. In wealthy countries like the United States, it allows commerce to function without even needing a break, thanks to reliable and well-maintained electricity generation and delivery systems, described further below.

4.3.1 ELECTRICITY GENERATION

In the book *Understanding Electric Power Systems* by Casazza and Delea (2010), electricity generation is described as "the process of conversion of energy resources to electric energy in order to be able to reliably supply customer requirements at all times." Electricity generating plants vary widely to meet customer needs and to interface properly with the primary energy source being exploited in a particular locale. As such, the energy conversion technologies (those that input a fuel or other source and output electrical energy) also vary.

Thermal power plants convert fossil, biomass, or nuclear fuels into heat (via combustion or fission) in order to convert water into steam, to then drive large turbine-generator systems to produce electricity. This method of thermo-mechanical energy conversion remains the dominant method for producing electricity in the world (EIA 2012). For established renewable energy sources like hydro and increasingly important sources like wind, the need for steam is not necessary—turbine-generator systems are placed where water and wind flow, thus converting mechanical energy into electricity directly.

Additionally, electricity produced at a carbon-based thermal power plant will tend to have large plant losses, in the form of waste heat released from the stack, or released from cooling water. It is typical for one unit of fuel input at such a plant to result in somewhere between 0.33 and 0.4 units of electric power output, and 0.6–0.67 units of produced waste heat. At a renewable power plant, on the other hand, it is common to *assume* only negligible losses. Thus, at a hydropower plant, for example, one unit of input power from water will result in one unit of output electric power. As described in *Energy Systems and Sustainability*, quantifying fuel use in the former process tends to be a bit more straightforward than in the latter (Boyle et al. 2003). We can measure the amount of coal, oil, or gas used in the production of power at conventional plants fairly accurately, but we would be challenged to measure accurately the amount of wind, water, or solar energy that passes through a turbine or other electrical array producing equivalent power.

Thus, when planning a new power plant, a number of factors impact the type and size of generator selected. The choice is mostly driven by the daily regional load cycle, and by cost—both installed and operating. For example, a generator that will be used to supply base load power needs must be able to operate for extended periods with a reliable and continuous source of fuel. Gas, coal, and large, dam-based hydropower are good candidates for base load generation and will tend to have higher installation costs and lower operational costs (Casazza and Delea 2010). Also, the application of some new technologies, especially co-generation of heat and power can significantly increase the overall efficiency of a thermal power plant by tapping a portion of the waste heat for space or process heating needs. This may increase the installed costs of the plant, but over the long term, will be balanced by the savings associated with using less primary fuel.

4.3.2 ELECTRICITY DELIVERY

Once electric power is produced, it must be transported to other demand sectors. The process of transporting electricity is depicted in Figure 4.1. First, the power plant generates electricity, where it travels to transmission substations. Next, the electricity passes through a voltage step-up transformer to prepare it for long distance travel to the demand sectors. As the electricity approaches its final destination, it passes through a power substation, then a voltage step-down transformer to deliver lower voltage electricity via distribution lines. It is stepped down again through a transformer drum on "power poles" to deliver safer, low voltage electricity to the customer (i.e., a home or a business).

Getting electricity from the power plant to the customer is done every day in the Industrialized World with what appears to be relative ease, though in reality, it is more complicated

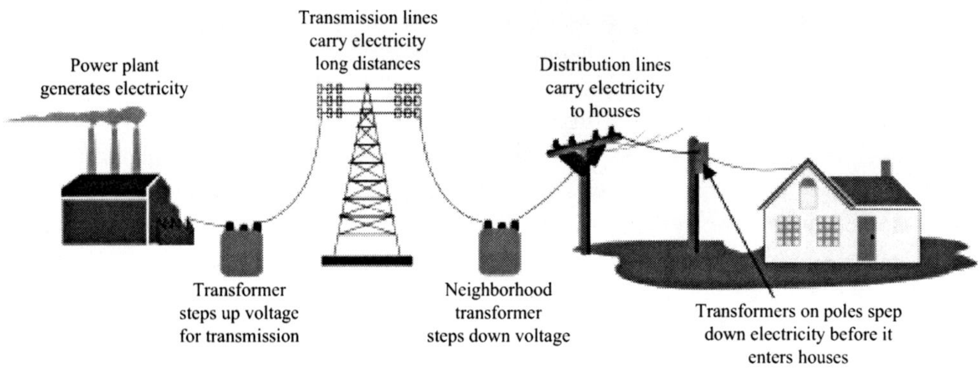

Figure 4.1. The process of transporting electricity.
Source: eia.gov

than it appears. A substantial issue with electricity is that large quantities of it cannot be stored effectively and economically, and thus must be produced at the same time as it is consumed (EIA 2012). This means that in traditional, centralized electricity delivery systems, power plants must be producing continuously and operating "on-demand," meaning that there exists the potential for a large degree of electricity "waste."

4.3.3 ENERGY CONSUMPTION IN THE ELECTRIC POWER SECTOR

The EIA produces a number of useful energy flow diagrams such as the one for electricity flow depicted below in Figure 4.2. This depiction is helpful for describing, in general, energy consumption in an electric power sector.

In the American example, the United States electric power sector consumed 40.04 Quads (42.2 exajoules) of primary energy. Of this total, 66% came from fossil fuels, 20.6% came from nuclear electric power, and 12.8% came from renewable energy. Most of the energy consumed in the sector (63%) was lost during the conversion of primary energy into gross generation of electricity, which amounted to 14.82 Quads. An additional 2% was lost from power plant use, amounting to a net generation of electricity of 14.01 Quads. It is this electricity that was transmitted and distributed to the other economic sectors. Using typical electrical energy units, this amount translates to a bit less than 4 TWh (3,856 billion kWh) of electricity consumption in the US economy.

So, how does this compare to consumption in other regions? According to the United States Central Intelligence Agency's *World Factbook* (2012), the top three electricity consuming regions in about this time period were China (4.7 TWh consumed in 2011), the United States (3.86 TWh in 2011), and the European Union (3.04 TWh in 2009).

4.4 THE TRANSPORTATION SECTOR

Mobility has been essential to the development of affluent economies. The very concept of competitive advantage in economic theory is predicated on the ability to move goods from producer to consumer. Without mobility of goods, nations and people are largely relegated to

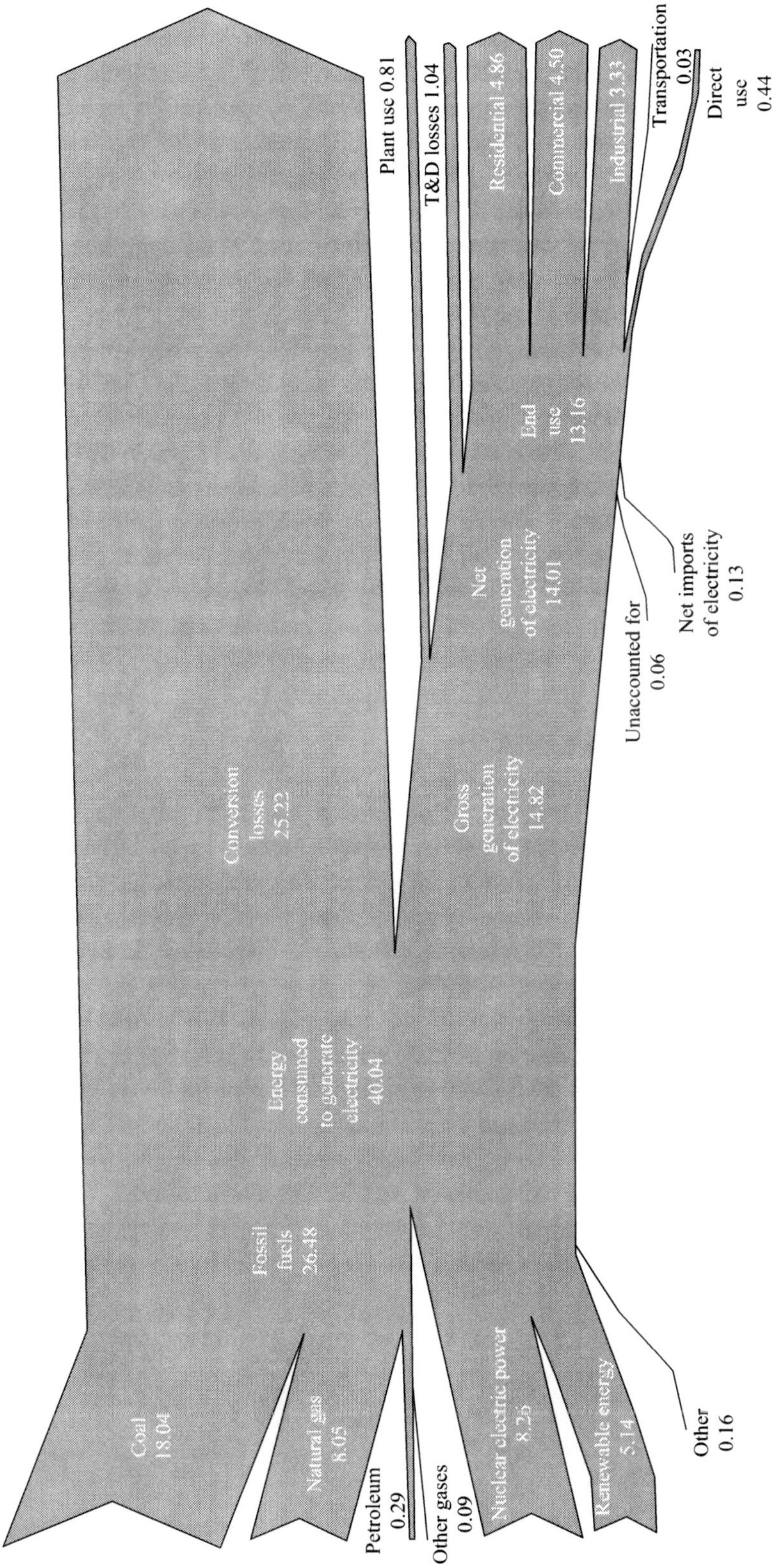

Figure 4.2. United States electricity flow.
Source: EIA Annual Energy Review 2011.

subsistence economies. Goods are produced all over the world (in locations that enjoy competitive advantage in raw materials or labor) and shipped to consumers, often traveling thousands of miles in the process. This is only viable in the case that energy is both abundant and cheap.

Mechanized transportation served to open vast tracts of land for settlement, none more evident (or more dramatically) than in North America. The settlement of large, sparsely populated spaces after the advent of mechanized travel allowed communities to spread broadly, with attendant benefits and costs. It has made the world a smaller place, in which it is possible for many people to experience a wide variety of places and cultures. At the same time, it has accelerated energy consumption dramatically, added hours each day commuting for many people, and, arguably, disrupted our sense of community.

The individual car or automobile has come to be one of the major consumers of energy in the world, while serving to transform affluent societies remarkably. In his well-received book *Home from Nowhere*, James Howard Kunstler notes, "By the mid-twentieth century, owning a car had become a prerequisite for first-class citizenship in the United States. It was assumed that every adult would own a car and use it for the most mundane tasks—to go to work, to get lunch, to buy aspirin..." (Kunstler 1996). It provides amazingly flexible transportation, which can take the driver anywhere there are roads (and sometimes beyond). In recent years, the Sport Utility Vehicle, popularly known as the SUV, although having many features of a truck, are commonly used for personal transportation. In whatever its form, personal transportation accounts for most of the petroleum use in the world's largest energy consumptive societies.

4.4.1 VEHICULAR TECHNOLOGY

The early automobile—a true breakthrough in personal transport—employed a variety of propulsion systems and fuels, including petroleum products, biofuel alcohol (ethanol), electricity, and steam. The ability to produce vast quantities of inexpensive gasoline and diesel fuel facilitated the triumph of the internal combustion engines (gasoline and diesel) in the early 20th century.

Internal combustion engines (ICEs) have significant advantages over the external combustion steam engines that predate them. Steam engine technology requires that a furnace boil water and then channel the resulting steam to drive an engine. This invariably requires the steam engine to be big and dependent on multiple stages (Boyle et al. 2003). The ICE, on the other hand, has the advantage that the fuel can be combusted directly in the engine, eliminating a very cumbersome step. It also introduced the potential for liquid fuels derived from alcohol or petroleum to be used inside the engines. These highly transportable liquids, with high-energy content, afforded the revolution in transportation we take for granted today.

Today, the ICE dominates transportation, but as humanity moves forward, we envision five basic designs will continue to shape or reshape how we convert fuels to power transportation:

- Gasoline engines
- Diesel engines
- Hybrid electric systems
- Electric vehicle systems
- Fuel cell systems

Gasoline engines are an ICE design that work on the premise that if you place an energy-dense fuel into a closed chamber and ignite it, much of the fuel's embedded chemical energy will be released via an explosive gaseous-phase combustion expansion, making it available to do useful work. If the engine design is such that it establishes a continuous cycle of combustion expansions over short time intervals, it can then be converted to mechanical work such as driving a reciprocating piston (Ferguson and Kirkpatrick 2000).

Modern gasoline engines use a four-stroke combustion cycle, also known as the Otto cycle, to convert the energy from gasoline fuel into vehicular motion. The four strokes in the cycle include intake, compression, combustion, and, exhaust. Pistons, placed in the appropriate configurations inside of a chamber, will move from one end to the other, allowing fuel–air mixtures to enter the engine where they are compressed. A properly timed spark is then used to ignite the gasoline, causing an explosion, which drives the piston back in to release any unused gases through an exhaust as the engine prepares for a new cycle to begin. These reciprocal movements are converted into rotation by connecting the pistons to a crankshaft, which is useful for turning the wheels of an automobile, for example. While overall efficiency is relatively low, the technology is effective and reliable, and with the availability of abundant and cheap petroleum, efficiency has not been an issue for much of its use.

The diesel engine cycle is distinguished from the gasoline cycle in its ability to ignite a fuel–air mixture without a spark. That is, in the diesel cycle you compress the air, and then you inject the fuel. The heat generated from the compression is sufficient to ignite the fuel. This allows for a high compression ratio, which yields greater engine efficiencies and, thus, less fuel used to accomplish equivalent mechanical work. It also allows for higher air compression ratios, which can provide more power than a gasoline engine under certain conditions (EERE 2003). Diesel engines are widely used in many vehicular applications. Perceptions of environmental issues have impacted widespread use in the United States for private vehicles, but they are the dominant engine choice in trucks. In Europe, they are widely used across the board.

Hybrid gasoline-electric (or diesel-electric) systems are used to improve the efficiency of the ICE even further, while also reducing tailpipe emissions, by mating a gasoline or diesel engine with a battery. In general, hybrid electric systems are comprised of a fuel tank, an engine, a battery pack, an electric motor, a controller, a generator, and a transmission (see Ehsani et al. 2009). These components interact with the vehicle's wheels in a few basic configurations. In a series configuration, the engine is used only to power a generator, which in turn charges the battery pack. The battery pack, in turn, drives an electric motor that is used to turn the transmission. In a parallel configuration, the engine can both turn the transmission and power the generator, which itself can charge the battery pack and turn the transmission (Figure 4.3). Batteries can also be partially recharged by recovering wasted energy from the vehicle's braking system. Hybrid electric vehicles can be used in automobiles, buses, or in locomotives (in a slightly different process).

Electric vehicles (EVs) are envisioned as the technology that could transition society away from ICE and hybrid electric technologies for automotive needs, and now are available in most affluent markets. The appeal of the EV is that there are no tailpipe emissions. Moreover, if the electricity could be sourced from solar, wind, or hydropower, then the vehicle would be free from fossil fuel energy (though petroleum would still be needed for the tires and plastic that make up the car). To be clear, this is the appeal, but in reality, the electricity currently comes from whatever primary energy is available locally, usually not from the sun or the wind.

With respect to EV design, it is quite simple: an electric motor, which is powered by a controller, which is in turn powered by electricity stored in a rechargeable battery pack, drives the wheels of a vehicle. The larger the battery pack, or rather the larger the battery pack capacity, the

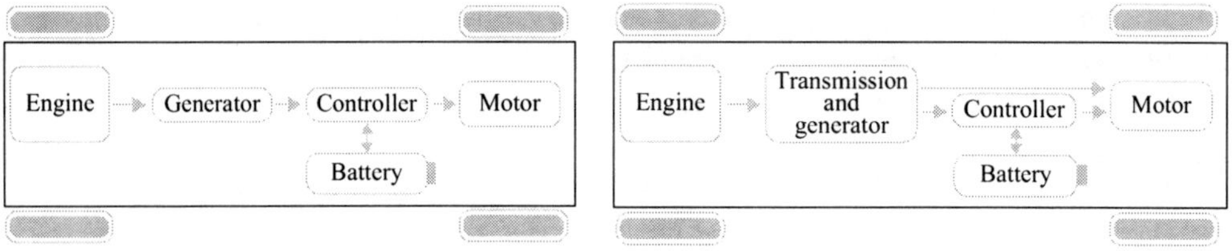

Figure 4.3. Hybrid electric system designs (left: series hybrid; right: parallel hybrid).

farther the vehicle will travel before needing to be recharged. This is really the only limitation of the technology. The problem is that this has turned out to be the most important limitation to allow EVs to penetrate the personal transport market, because battery packs are expensive and the driving range between recharging is roughly 100 miles (161 km), a range unacceptable to a society used to being able to travel longer distances and with reliable fueling infrastructures. Further, it is not yet clear how electricity grid systems will be affected by EV uptake, though this issue is being investigated quite extensively (Lampropoulos et al. 2010).

Finally, hydrogen fuel cell vehicles (FCVs) have been spoken of as a technically viable ICE replacement option since before the 21st century. The appeal of this technology is that the only tailpipe emission is water, and that the energy carrier, hydrogen, is by far the most abundant element in the universe. The problem is that hydrogen is not really available in elemental form on earth and needs to be extracted from other compounds. The hope is that one day water atoms can be efficiently broken apart via electrolysis (or other processes) using solar, wind, or geothermal energy as the electricity source to create the hydrogen needed to operate the FCV (NREL 2007; EERE 2012). The reality is that natural gas will be "cracked" to provide the hydrogen fuel for FCVs for most of the world for the foreseeable future.

As for FCV design, the most likely technology that will be used for vehicular application is the proton exchange membrane fuel cell (PEMFC). PEMFCs consist of a proton exchange membrane and a catalyst, sandwiched between an anode and a cathode. The role of the PEMFC is to convert hydrogen and oxygen (usually from air) into water, in an effort to produce electricity. After that, the FCV is essentially an EV, but with one potential advantage: as long as there is enough hydrogen on-board, the range can be extended to what is deemed acceptable by most markets (300 miles or 483 km). Obstacles that FCVs face in terms of potential market penetration include extremely high costs, difficulty storing hydrogen gas abundantly and safely, and a total lack of refueling infrastructure. While there are many reasons to want to see hydrogen FCVs succeed, it is likely that the greatest potential for fuel cells in the short term is in stationary power applications (EERE 2012).

4.4.2 AUTOMOBILES VERSUS MASS TRANSIT

Whether in the United States or in Moçambique, transportation is not only a highly energy-intensive sector, but it is also absolutely essential to economic development. The United States constitutes a special case due to the vast spaces that colonists found to expand into as they almost completely displaced the native populace. The colonization coincided somewhat with the advent of rail and of steamboat transportation. Highways largely displaced rail transportation in the latter half of the 20th century in the United States, though. The extensive surface transportation infrastructure permitted the American population to spread out and use more of the large land mass for living.

It converted vast portions of agricultural land into residential areas and allowed people to live tens of miles from their work. This had a role in spurring prosperity as suburbs and then exurbs provided space for the average American to own a home—frequently on a piece of land that emulates a European estate, with sprawling lawns. There are significant costs incurred by the lifestyle of sprawl, but it put more wealth in flux and supported the growth of a huge Middle Class. In *Home from Nowhere*, Jim Kunstler paints a colorful picture of how America grew in the wake of the Second World War and suburbs emerged as a dominant living form. He critiques this form aesthetically as well as practically, observing, "We are going to have to develop a different notion of the good life and create a physical form that accommodates it."

The sprawl has also been associated with inefficient transportation in individual vehicles. The costs are large, but easy to ignore on a day-to-day basis. Building and maintaining the massive infrastructure to support some quarter billion cars in the United States is tremendously expensive. The cost to human lives associated with the personal vehicle lifestyle is huge as well: averaging in the neighborhood of 40,000 fatalities per year in the United States. The data at the time of this writing indicate a decline to a recent record low of just under 34,000 fatalities—partially explained by the decreased driving during the 2007–2008 recession (NHTSA 2010). That figure does not include the much larger number of people maimed or seriously injured. Nor does it consider the cost to animal life and ecosystems of crisscrossing the landscape with highways. The lifestyle has given us unprecedented mobility, with an associated sense of freedom. The costs of this lifestyle are likely to become apparent to the typical consumer as energy prices rise—and material prices rise along with them, as energy is required in all kinds of manufacturing—and even road maintenance.

Mass transit is an alternative form of energy consumption in the transport sector. It consists of publicly available transportation services paid, in part, by revenues from taxes and usage fees. Common examples include bus services, light rail, subways, and ferries. (Airlines may also be considered mass transit.) The availability of land—assuming relatively low fuel prices—is the chief determinant in whether mass transit will be the dominant transportation mode over the private automobile. Also, once one mode has gained sufficient traction and infrastructures are in place to support that choice, it can be very difficult for alternative modes to become viable choices. The freedom afforded us by cheap fuel and cheap cars may only be changed when neither the fuel nor the steel (or other raw materials[1]) are cheap enough to merit this mode of transportation. We suggest that conservation will be an important part—reducing our demand for transportation will make the transition much easier. If we live close to our work, we not only eliminate significant transportation miles but also increase the viability of public transportation, as population centers become denser again. This is particularly true for America. Furthermore, the title of this section is itself potentially misleading: public transportation does not oppose private vehicles. Indeed, effective public transportation, which people opt to use, reduces congestion on the surface roads, allowing private vehicles to be more effective and pleasant.

4.4.3 COMMERCIAL TRANSPORTATION

Moving goods is perhaps more essential to the economy than is moving people. Consumer societies dominate much of the Industrial World. They need goods produced from around the world. It has allowed the import and export of all kinds of food from all around the world, year round. It has also allowed the export of pollution and other externalities to countries that are now

[1] Note that lightweight plastics and other synthetics are petroleum products, so these materials are also likely to be in short supply when there is a petroleum shortage.

producing more of the raw materials and intermediate products (e.g., steel and aluminum). The most affluent nations are relieved of much of these productive activities through the tremendous use of energy for global transport. Of course, a great deal of product is transported within the consumer nations' boundaries as well.

Public or mass transportation of people fits in the category of commercial transportation as well. Buses, trains, and airplanes carry passengers as well as cargo. In general, mass transportation is likely to be safer and more efficient, in terms of both resources and time. Nevertheless, it is much less popular in the United States than individual vehicular transportation. The population densities around urban centers that grew up in Europe and Asia prior to the advent of mechanized transportation are conducive to effective public transportation, while the sprawl of urban and exurban areas in the United States make it very difficult.

Excluding the energy needed upstream in the vehicle manufacturing process, the transportation sector is most sensitive to the price of oil, since it is almost exclusively dependent on gasoline and diesel. Whether automobiles or airplanes or interstate rail, very high energy densities are required so that motive technologies will carry the vehicles and its passengers or cargo hundreds or even thousands of miles. Petroleum products stand well apart in this category. For example, one kilogram of gasoline contains as much energy as one hundred kilograms of lithium-ion batteries. Furthermore, the fuel can be replaced thousands of times at minimal cost, without degrading the overall system, whereas batteries can only be recharged a finite number of times before requiring complete replacement. The energy density of petroleum products provides us with the freedom to travel where we wish, when we wish. It liberates us from routes and schedules. But it ties us to one commodity—a commodity which will almost certainly be in tight supply by the middle of the 21st century.

4.4.4 *ENERGY CONSUMPTION IN THE TRANSPORTATION SECTOR*

The *Transportation Energy Data Book* classifies energy consumption in the transportation sector by transportation mode and fuel type (CTA 2012). Highway modes of transportation include light vehicles (cars, light trucks, motorcycles), buses, and medium/heavy trucks. Nonhighway vehicles include air (domestic and international), water (freight and recreation), pipeline, and rail (freight and passenger). In 2010, the United States consumed 27.6 Quads to satisfy transportation needs. About 81.7% was used by highway vehicles (58.8% by cars and light trucks, 22.3% by heavy trucks, and 0.69% by buses). The remainder came from nonhighway vehicles (7.77% by air, 5% by water, 3.4% by pipeline, and 2.1% by rail). Gasoline and diesel fuel provided the bulk of this energy. For reference, the EU percent distribution of transportation mode is not dramatically different from that in the United States (though aviation has grown dramatically since the inception of the EU); however, the total energy consumed is roughly half that of the United States at 14.5 Quads (Eurostat 2010). Finally, estimated transportation consumption in China is about six Quads and rapidly growing (IFEU 2008).

4.5 THE INDUSTRIAL SECTOR

Prior to the first extensive use of fossil fuels, firewood and animal power provided most of the energy input needed to satisfy society. As populations grew and advanced economically and socially, it became apparent that more intensive and more portable energy sources would be needed to support wide-scale mechanization. Wind and water power, and charcoal answered that challenge initially, but it was coal that permitted tremendous expansion of industry. The fact that coal had been accumulated and intensified over millions of years made it a source that dwarfed the biomass fuels (charcoal included), where the forests only represented a few hundred years of accumulation. By the year 1900, coal was providing as much energy to London as the sustainable harvesting of a forest six times the size of the British Isles (Ebenhack 1995). The Industrial Revolution could not have reached fruition without the abundant energy of coal.

Today, industries representing the manufacture of a number of goods vital to the functioning of modern economies continue to depend extensively on coal, as well as on oil, gas, and electricity. As was alluded to in Chapter 1 of this text, at a primary level, this modern industrial sector includes important activities like petroleum refining, and steel and aluminum processing, described in further detail below. Additionally, cement, paper, and chemicals industries round out the sector. These latter industries require refined petroleum products, steel, aluminum, electricity, and transportation to function. Indeed, every manufactured product of the 21st century contains a large amount of embedded energy. This truly is an unavoidable consequence of modern economic activity.

4.5.1 PETROLEUM REFINING

Crude oil is a robust and valuable resource, comprised of hundreds of different molecules containing hydrogen and carbon—appropriately named, hydrocarbons—that can be separated to produce thousands of useful consumables ranging from heating oil and jet fuel to medicines and crayons. The hydrocarbon separation process, accomplished by fractional distillation, or "fractionation," is at the heart of the petroleum refining industry (Leffler 2008).

Hydrocarbon molecules are distinguished from each other by their carbon–hydrogen chain lengths, with each chain length having different physical properties. Refineries take advantage of differences in the density and boiling point of the various molecules within petroleum by heating and vaporizing (or boiling) it and then condensing (or separating) out desired hydrocarbon "fractions" that can then be used for different applications. After fractionation is completed, refineries also treat each chain to remove any impurities that may be present. Additionally, modern refineries use secondary processing to break longer chains into shorter chains; to combine shorter chains to make longer chains; or, to alter chains to make desired hydrocarbons—all at a particular fraction (EIA 2013).

In Figure 4.4, we see a simplified depiction of the petroleum fractionation process. Clearly, the petroleum refining industry provides inputs to virtually every economic sector, including the transportation sector, the electric power sector (especially for backup capacity in much of the African continent), and the chemical industry. All of the steps involved in the refining process—distillation, chemical processing, impurity removal, and pollution abatement—require energy to complete. Indeed, it is estimated that operating costs tied to energy needs can range anywhere from 40%–50% in modern plants (NEED 2010). The fractionation process gives society access to a multitude of fuels and goods that we (currently) cannot live without.

4.5.2 THE STEEL AND ALUMINUM INDUSTRIES

Steel is extremely strong, is extremely malleable, and can withstand heat much greater than most other substances used in society, both past and present. Comprised almost entirely of iron

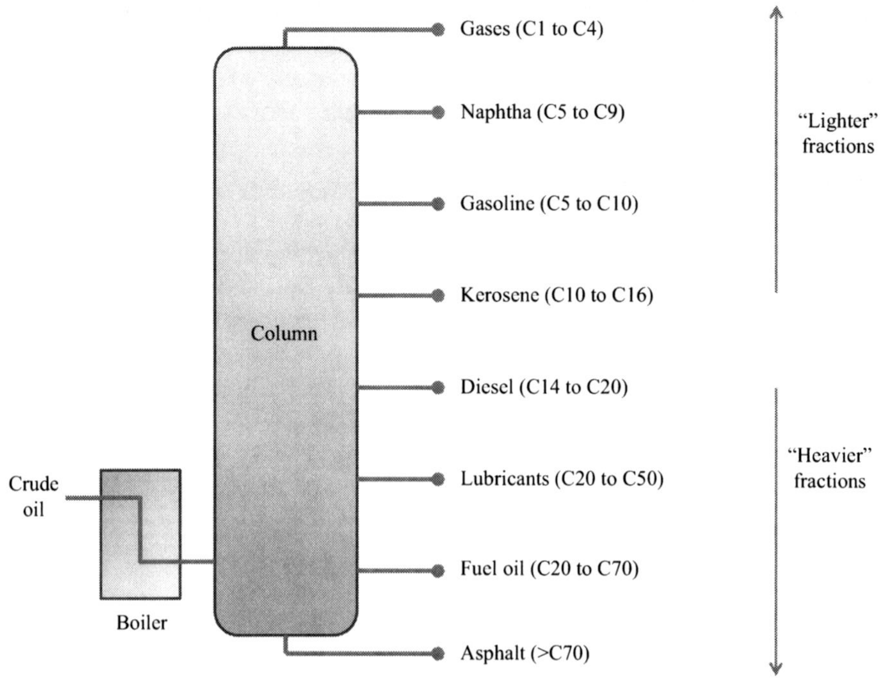

Figure 4.4. A simplified view of the petroleum fractionation process. Less dense, lower boiling point carbon chains boil off as "lighter" fractions, while denser, higher boiling point carbon chains boil off "heavier" fractions.

and carbon, it is also extremely abundant. According to the World Steel Association (WSA), the manufacture of primary steel (directly from iron ores) involves a multistage process, with each stage requiring an impressive amount of energy, whose cost constitutes anywhere from 20% to 40% of the total cost of its production (WSA 2008). Essentially, iron-rich sedimentary rocks are mined, transported, and ground into powder, where iron ore gets separated out with powerful magnets. Traditionally, pure carbon fuel (i.e., coke) is then combined with this ore in a blast furnace to make molten liquid iron at temperatures in excess of 2,700°F (1,480°C). Steel is then formed by adding recycled, secondary steel scraps to the furnace and removing impurities. Secondary steel can be manufactured directly using electricity in electric arc furnaces. Primary steel production accounts for about 75% of world steel production, with secondary steel production accounting for the rest.

Aluminum is another important metal used in the world's economy. It is widely used because it is both light and strong, and it is resistant to corrosion (Emsley 2011). Because of this, aluminum is used quite extensively in transportation, packaging, and building and construction industries. As with steel, aluminum is derived from earth-mined ores, primarily bauxite. Bauxite is processed to produce aluminum oxide, which then requires the separation of aluminum and oxygen via electrolysis in an electric arc furnace. In all, it takes roughly four tons of bauxite to produce one ton of aluminum (Alcoa 2012). The energy intensity of this process has been reduced by mixing new aluminum with recycled aluminum, which accounts for a growing portion of new production.

Aluminum and steel are two materials integral to modern society. Some of the products made from these manufactured substances include cars, planes, trains, bridges, tools, buildings, gutters, food and beverage containers, food and medicine additives, foil wraps, light bulbs, pots and pans, solar panels, and wind turbines. These metals are essential building blocks to the world's economy, as is the energy required to produce them.

4.5.3 ENERGY CONSUMPTION IN THE INDUSTRIAL SECTOR

As per EIA reporting, primary fuels in the industrial sector are consumed both for energy and for chemical feedstock (EIA 2012). In terms of energy consumption in the United States, natural gas and the energy carrier electricity are the predominant energy "sources" in the sector, though wood and wood waste also represent a significant portion. In total, the United States industrial sector consumed about 20.4 Quads in 2011 (21% of total consumption). And of those 20.4 Quads, the petroleum refining industry consumed the most (32%). This was followed by the chemicals industry (24%), paper industry (11%), and metals industry (7%) (EIA 2006). We note that the low percentage represented by the metals industry is a bit deceptive with respect to the importance they play in modern economies, since much of the world's steel manufacturing now occurs in China. (Thus, energy consumption and emissions in the United States and EU have been exported.) Indeed, the industrial sector represents the majority of that country's energy consumption (about 74% in 2008).

4.6 THE RESIDENTIAL AND COMMERCIAL SECTORS

The residential and commercial sectors represent those parts of the economy that intersect most with the day-to-day endeavors of average citizens, both for living and for working. And built structures represent perhaps the most palpable and direct energy consuming elements of these sectors. They are expected to provide light, heat, running water, and electricity to operate appliances for refrigeration, communication, and various forms of entertainment. As was noted in Chapter 1, lighting and heating systems can remain in place for several years, while communication, computation, and entertainment systems can be replaced in just a few years. The book *Residential Energy* by Krigger and Dorsi (2009) does an excellent job discussing this subject matter, and we will try to highlight and expand on some of their most relevant points in the subsections below.

4.6.1 LIGHTING

From windows to candles to electric lamps, different technologies have been created throughout time to achieve the appropriate amount of lighting to complete specific human tasks. Light output and illumination are important factors when choosing a lighting option, and electric lamps are well suited to meet these criteria in residential and commercial applications. Three notable electric lighting system technologies employed in the residential and commercial sectors include traditional incandescent lamps, increasingly popular fluorescent lamps, and the relatively new light emitting diode (LED) lamps.

Incandescent lamps are the most commonly used lamps. The technology—resistive heating of a tungsten filament to produce a white-hot "glow"—has not changed substantially since its introduction by Thomas Edison in 1879. It is simple and cheap, and thus, has been widely used for lighting since its inception, especially for residential applications. Its primary drawback is that it is inefficient as a lighting system, producing a substantial amount of wasted heat.

An alternative to the incandescent lamp is the fluorescent lamp. The fluorescent "glow" comes primarily from the excitation of mercury vapor within the lamp. Fluorescent lighting systems, including the ballasts that are used to safely run them, consume about 75% less energy than incandescent systems with equivalent light output. At the residential level, a large movement has been initiated to relamp incandescent lamps with compact fluorescent lamps (CFLs) that can reduce overall energy consumption while utilizing existing incandescent light fixtures. It is believed that relamping efforts will lead to reduced electricity consumption; however some recent studies indicate that this may not necessarily be true if the efficiency savings are replaced by increased consumption due to the so-called "rebound effect" (European Commission 2011).

Finally, LEDs are an expanding lamp technology with extremely high efficiencies. LEDs work like photovoltaic cells, but in reverse. Rather than light hitting a semiconductor and mobilizing electrons, in the LED an electric current on a semiconductor produces luminescence (Energy Star 2012). LEDs emit light far more efficiently than both incandescent lamps and CFLs, with less waste heat than other technologies. This is primarily because LEDs emit light—and heat—in one direction, as opposed to the former technologies, which emit in all directions, thus wasting unnecessary light. A depiction of how heat is dissipated in modern lamps is available in Figure 4.5.

At the time of this writing, the cost of LED production is still high, compared to other technologies, but LEDs appear poised to take over a much larger share of the lighting market. Unlike most currently used fluorescents, LEDs are dimmable and they can last much longer than either incandescent or even fluorescent lamps. The main obstacle at this time is in the design of lamps that produce stable white light for general lighting applications, which remains a significant challenge. They continue to gain broad general retail market penetration.

4.6.2 *HEATING*

The dominant heating practice remains to be the direct combustion of a fuel for most of the world. This technology dates back to some of humanity's earliest uses of fires. An open fire is an extraordinarily inefficient means of space heating, likely much less than 10% efficient.

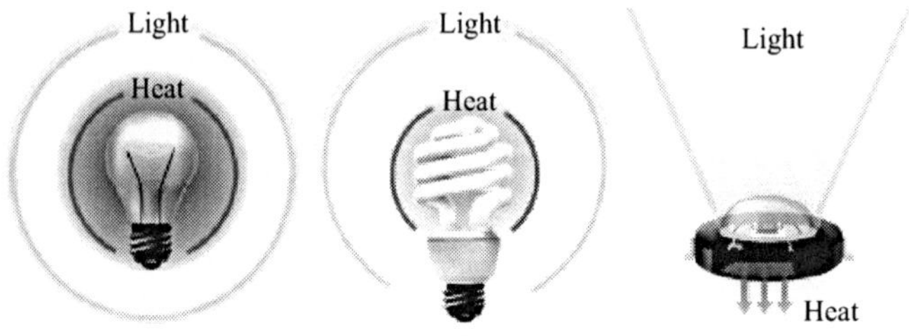

Figure 4.5. Directional light and heat emissions from incandescent, compact fluorescent, and light emitting diode lamps.
Source: energystar.gov

Once the fire is enclosed in a fireplace, some of the heat is transferred to the stone or brick and reradiated out toward the living area. A well-designed fireplace and chimney can be plausibly efficient. They can be made more efficient by incorporating a fan, which provides forced convection, pushing heated air out into the living space.

But as time passed and as the fossil fuels became more available, it became more common in built structures to employ centralized heating systems that take advantage of the higher energy content of the fuel and of the special properties of water, specifically latent heat and heat capacity. In these systems, a fossil fuel burner would be used to heat water so that it could then circulate through steam radiators or hot water baseboard heaters located around a building. These methods provide sufficient and controllable heat to warm a building. Water-based heating systems continue to be used, especially in older housing stock and in some institutional buildings. Furnaces provide even more efficient combustion chambers and heat exchangers to transfer as much of the heat as possible to air being circulated through a building, through ducted, forced convection. Modern natural gas-fueled furnaces can achieve nearly 95% efficiency in extracting the chemical energy content of the gas and delivering it to the living space (Krigger and Dorsi 2009). Finally, the broad dissemination of electricity has led to the use of electrical resistance to provide space heat. Much like hot water baseboard heating, electrical heaters can be placed around a building, using electricity to generate the heat. At the point of use, electric heating is easily controlled and emission free. However, the total energy fuel cycle can be rather inefficient and dirty, since the electricity is typically generated at power plants that utilize coal or natural gas as primary energy sources.

4.6.3 COOLING

Cooling needs are almost entirely based on location, design of the structure, and how well the occupants can tolerate heat (Krigger and Dorsi 2009). Thus, in an ideal scenario, a building would utilize landscaping and proper ventilation to reduce the amount of solar gain on the structure before looking to employ artificial cooling such as using air-conditioning. The reality is air conditioning is usually the option people choose first.

Air conditioning essentially is based on the same principles as refrigeration. In a refrigeration cycle, the point is to absorb heat from a space you want to cool, like the inside of a refrigerator or the air passing through an air conditioner. The heat vaporizes a working fluid, such as (the now largely banned) Freon or ammonia. The vapor is then condensed by compression and heat is rejected outside of the environment being cooled: the kitchen or the outdoors. Thus, if one were to open the door to a refrigerator in order to cool a room, as people have been known to do, they will actually add heat to the room, because all of the heat taken up by vaporizing the refrigerant goes into the room, as well as waste heat from the compressor/condenser and pump. Some forms of heat pumps work on the same principle, but its circulation is reversible, so that the working fluid is vaporized by the ambient heat in the outdoors and released into the living space during the winter, when the vaporized fluid is compressed to the liquid phase in the living space.

4.6.4 APPLIANCES

Beyond space heating and cooling and lighting, we ask our dwellings to help us perform a number of domestic activities that require energy, and there are a number of appliances that we bring into the home to do various tasks. The most energy intensive appliances include refrigerators and freezers, dishwashers, clothes washers, and dryers. Refrigerators and freezers are primarily used in the home to extend the life of perishable foods. These appliances do not necessarily draw much electrical power but because they are kept on indefinitely, they end up representing a large percentage of electricity consumption in the home. Computers, insulation, and other design improvements have dramatically reduced the energy needs of refrigerators compared to their predecessors (Krigger and Dorsi 2009), which has allowed for the manufacture of larger units, thus reducing efficiency gains. However, the largest energy losses now come from the way individuals use these appliances. Dishwashers, on the other hand, are both energy and water intensive, but most of their consumption is indirect, depending on the home's heated water tank to perform its dish cleaning duties. Clothes washers are also energy and water intensive but they do not necessarily have to use heated water to effectively clean clothes. In fact, doing laundry in cold water routinely can save a significant amount of energy. Almost gone are the days of line-drying laundry in much of the affluent world, especially the United States. Now, mechanical clothes dryers are routinely employed. Powered by gas or electricity, dryers dominate residential buildings and can be extremely energy intensive.

4.6.5 CONSUMER ELECTRONICS

Because of the amount of energy consumed by these devices and their importance in virtually every facet of life in the Industrialized World, we feel compelled to mention briefly the dramatic uptake of consumer electronics in the past few decades. In 1980, the average United States home had about three electronic devices, and according to the Consumer Electronics Association, this number has multiplied eight times to 25 devices 30 years later (Wall Street Journal 2012). Roughly, 7%–15% of all electricity usage in a home now comes from consumer products like TVs, PCs, stereos, video game consoles, mobile phones, and tablets. In commercial settings, it is not uncommon to have PCs and monitors left on when not in use. Thus, consumer attitudes are largely to blame for the dramatic increase in energy use, with few people adjusting energy management settings that are preprogrammed in most modern electronic devices to reduce consumption, or by shutting down the devices altogether when not being used at all.

Consumer electronics are finding their ways increasingly into lower income societies as well. It is remarkable to note that in some African nations more people have cell phone connections than have electricity. How can this be? Charging stations have proliferated, at which people can pay to charge their phones. While visiting colleagues in Moçambique, one of them purchased cell phone time from a vendor and then turned to remark on how seductive it was. Although he was saving money to build a house, the cell phone time was sold in such small units, that it was an easy, spontaneous act to buy time, permitting more cell phone use.

4.6.6 ENERGY CONSUMPTION IN THE RESIDENTIAL/COMMERCIAL SECTORS

In 2011, the United States residential and commercial sectors accounted for 10.6 Quads of energy consumption, or 10.9% of the United States total (EIA 2012). Natural gas and petroleum represented the largest energy sources used by these sectors, mostly for space heating. Electricity was the next most used form of energy. As in the other sectors, Europe's distribution was comparable, but with smaller total energy use, and with a higher reliance on biomass for heating. In China, coal remains an important end-use heating choice, however natural gas has been gaining market share as coal production has been heavily redirected to electricity production (LBNL 2009).

4.7 IMPROVING ENERGY EFFICIENCY IN ECONOMIC SECTORS

As we close out this chapter on economic sector energy consumption, we note that over the years, technological advances, fuel switching, "outsourcing," and for a while, conservation, together have greatly reduced energy consumption in virtually all of the economic sectors of the Industrialized World. This is especially true since the energy crises of the 1970s. In the transportation sector, we now have more stringent corporate average fuel economy standards that are beginning to result in marked increases in the efficiency of cars and light trucks. We now have extremely efficient lighting options. Appliance energy efficiency has increased tremendously. The industrial sector's energy efficiency has improved significantly, thanks in part to the deliberate recycling of raw materials. And in the residential and commercial sectors, more efficient and cleaner burning natural gas and electricity has almost entirely displaced coal use for heating. Yet all of these improvements have been balanced by ever-increasing consumption of primary energy, primarily from fossil fuels. Moreover, many of the most energy-intensive industries have been exported to Developing Nations, which has artificially enhanced the energy efficiency (and air quality) in affluent countries. We now live in an extremely interdependent and geoengineered world. The choices we make impact future energy-consumption patterns and society and the environment as a whole.

REFERENCES

Alcoa. 2012. "It All Starts from Dirt." Film. http://www.alcoa.com/global/en/news/alcoa_video.asp?videoid=23

Boyle, G., B. Everett, and J. Ramage. 2003. *Energy Systems and Sustainability*. New York: Oxford University Press.

Casazza, J. A., and F. Delea. 2010. *Understanding Electric Power Systems*. New York: Wiley-IEEE Press. doi: http://dx.doi.org/10.1002/9780470588475

CIA (Central Intelligence Agency). 2011. "World Fact Book." https://www.cia.gov/library/publications/the-world-factbook/

CTA (Center for Transportation Analysis). 2012. "Transportation Energy Data Book." http://cta.ornl.gov/data/chapter2.shtml

Ebenhack, B. W. 1995. *Nontechnical Guide to Energy Resources*. Tulsa: PennWell.

EERE (Department of Energy Energy Efficiency and Renewable Energy Program). 2003. "Just the Basics: Diesel Engine." http://www1.eere.energy.gov/vehiclesandfuels/pdfs/basics/jtb_diesel_engine.pdf

EERE (Department of Energy Energy Efficiency and Renewable Energy Program). 2012. "Electrolytic Processes." http://www1.eere.energy.gov/hydrogenandfuelcells/production/electro_processes.html

EERE (Department of Energy Energy Efficiency and Renewable Energy Program). 2012. "Stationary/Distributed Generation Projects." http://www1.eere.energy.gov/hydrogenandfuelcells/tech_validation/generation_proj.html

Ehsani, M., Y. Gao, and A. Emadi. 2009. *Modern Electric, Hybrid Electric, and Fuel Cell Vehicles: Fundamentals, Theory, and Design*. New York: CRC Press.

EIA (Energy Information Agency). 2006. "Manufacturing Energy Consumption Survey." http://www.eia.gov/emeu/mecs/

EIA (Energy Information Agency). 2011. *Annual Energy Review*. http://www.eia.gov/totalenergy/data/annual/index.cfm

EIA (Energy Information Agency). 2012. "Energy Explained." http://www.eia.gov/energyexplained/index.cfm?page=us_energy_use

EIA (Energy Information Agency). 2013. "What is an oil refinery?" http://www.eia.gov/energy_in_brief/article/refinery_processes.cfm

Emsley, J. 2011. *Nature's Building Blocks: an A-Z Guide to the Elements*. Oxford: Oxford University Press.

Energy Star. 2012. "Learn About LEDs." http://www.energystar.gov/index.cfm?c=lighting.pr_what_are

European Commission. 2011. "Addressing the Rebound Effect." http://ec.europa.eu/environment/eussd/pdf/rebound_effect_report.pdf

Eurostat (European Commission Statistics). 2010. "Statistics Explained." http://epp.eurostat.ec.europa.eu/statistics_explained/index.php/Transport_energy_consumption_and_emissions

Energy Star. 2012. "Learn about LEDs." http://www.energystar.gov/index.cfm?c =lighting.pr_what_are

Ferguson, C. R., and A. T. Kirkpatrick. 2000. *Internal Combustion Engines: Applied Thermosciences*. Hoboken: Wiley.

IFEU (Institute for Energy and Environmental Research Heidelberg). 2008. "Transport in China." http://www.kfwentwicklungsbank.de/ebank/EN_Home/Sectors/Transport/Transport_in_China.pdf.

Kunstler, J. H. 1996. *Home from Nowhere*. New York: Simon and Schuster.

Krigger, J., and C. Dorsi. 2009. *Residential Energy: Cost Savings and Comfort for Existing Buildings*. Upper Saddle River: Prentice Hall.

Lampropoulos, I., G. M. A. Vanalme, and W. L. Kling. 2010. "A methodology for modeling the behaviour of electricity prosumers within the smart grid." *Proc. of IEEE PES Conference on Innovative Smart Grid Technologies Europe*.

Leffler, W. L. 2008. *Petroleum Refining In Nontechnical Language*. Tulsa: PennWell.

LBNL (Lawrence Berkeley National Laboratory). 2009. "China's Coal: Demand, Constraints, and Externalities." Report: LBNL-2334E. http://china.lbl.gov/publications/china%E2%80%99s-coal-demand-constraints-and-externalities

Lighting Research Center. 1994. "Lighting the Way: A Key to Independence." http://www.lrc.rpi.edu/programs/lightHealth/AARP/healthcare/lightingOlderAdults/tasklights.asp

NEED (National Energy Education Development Project). 2012. "Energy Consumption." http://www.need.org/needpdf/infobook_activities/IntInfo/ConsI.pdf

NHTSA (National Highway and Traffic Safety Administration). 2010. Report: DOT HS 811 346. http://www-nrd.nhtsa.dot.gov/Pubs/811346.pdf

NREL (National Renewable Energy Laboratory). 2007. "Potential for hydrogen production from key renewable resources in the United States." Report: NREL/TP-640-41134. www.nrel.gov/docs/fy07osti/41134.pdf

Smil, V. 2008. *Energy: A Beginner's Guide*. Oxford: Oneworld Publications.

Wall Street Journal. 2012. "Consumer Electronics Association Says Devices Getting Simpler." http://online.wsj.com/article/SB10001424052970204124204577149320992712052.html

WSA (World Steel Association). 2008. "Energy Fact Sheet." http://www.worldsteel.org/dms/internetDocumentList/fact-sheets/Fact-sheet_Energy/document/Fact%20sheet_Energy.pdf

CHAPTER 5

Petroleum and Other Energy Resource Limits

This chapter explores the role of resource limits on energy choices that will be made in a sustainable energy future.

5.1 EARTH'S ENERGY RESOURCE "BANK ACCOUNT"

Let us follow a popular practice of likening our energy use to personal finances. If we are fortunate enough to have some financial savings as well as financial income, then the income would represent a continuing financial *flux*, while the savings would be comparable to financial *reserves*. If we live only on our income, our lifestyle is *sustainable* as long as the income lasts. If, however, we are spending our savings, they will eventually disappear, thus making our lifestyle less and less sustainable. There are times that it makes sense to dip into savings, but we are on a fully unsustainable path if we use up savings to live. If we spend savings to build for the future (buy a home, invest in education, or a business in this analogy), then its use may support constructive economic growth.

As a thermodynamically open system, Earth receives a plentiful and continuous *flux* of valuable solar radiation to power ecological systems, allowing growth to be an essential component of our existence. However, much of modern humanity's economic/industrial growth has been based on tapping into energy *reserves* (first biomass, now fossil fuels) rather than simply using the solar flux of energy coming to Earth. As in the analogy above, relying on reserves to maintain the current global consumption system is beginning to come at odds with the known limits of nonrenewable, stock-based energy. It appears to be time to use these savings to build for an energy future that relies more heavily on flux-based energy.

In this chapter, we address the important issues of growth and limits, particularly how the most relied upon resource, petroleum, has been exploited and how its production will probably decline over the next several decades. We also address flux-based limitations described in Chapter 1 and how both stock and flux limitations will impact the goals set forth in a more sustainable energy pathway.

5.2 GROWTH AND LIMITS

Growth is a concept that has received wide attention with respect to sustainability. Even before the 1987 Brundtland Commission report on sustainable development, Dennis and Donella Meadows published a famous book called *Limits to Growth*, which focused attention on the issue that the rising demands for resources by a rapidly growing population cannot be met indefinitely with finite resource bases (Meadows 1972). *Limits to Growth* received a great deal of attention, including much criticism. The primary criticism had been that the issue of limits has been raised for centuries and the predictions of collapse have always failed. While this is true in recent centuries, many societies have indeed collapsed in the past, in part due to exceeding resource limits in the face of rapidly growing consumption. Why collapse has yet to be achieved in the modern era speaks more to the tremendous amount of solar capital stored inside of the fossil fuels than to humanity's miraculous ability to escape collapse through unbridled growth.

5.2.1 *THE GROWTH FUNCTION*

The growth that Dennis and Donella Meadows describe in *Limits to Growth* is that of exponential growth, which has turned out to be a simple, yet widely misunderstood mathematical concept (Bartlett 1979). In plain terms, an exponential function explains how a quantity, Q, grows in relation to itself over time, expressed mathematically as:

$$Q(t) = Q_0 e^{rt}$$

where Q_0 is the initial, starting quantity, r is the growth rate usually expressed as a percentage, and t is time (e is a numerical constant roughly equal to 2.718). Note the characteristic behavior of growth functions: they can begin quite small and seem to be growing rather slowly, but then begin to grow in value quite rapidly once the value builds some with time.

To make this concept simpler, let us focus on how long it takes Q_0 to double to $2Q_0$, or:

$$2Q_0 = Q_0 e^{rtd}$$

where td is the doubling time. Doing some manipulation, this doubling function reduces to

$$td \sim 70/r$$

This is known as the "rule of 70." That is, the time it takes to double a quantity growing at r percent is roughly equal to 70 divided by r percent.[1] That's really all you need to remember about exponential growth. Using the rule of 70, when you hear of something growing at 7% per year, it will take about 70/7 or 10 years to double.[2] Alternatively, if we think the human population will double in 35 years, then that means the annual population growth rate is 70/35 or 2%.

Speaking of population, let's apply this concept to recent population growth. As of the last decade of the 20th century, the world population growth rate was about 1.5% per year, which

[1] See Masters and Ela (2008) for a full derivation.

[2] The doubling function actually works out more precisely to a 'rule of 72', but one incurs a rather small error to think of it more simply as the rule of 70.

would equate to a 46 year doubling time. This rate should be seen as quite high within the context of historical human population growth. According to Hern (1999), by the time humanity approached the current era, the population had likely gone through doubling and halving several times before reaching a net doubling and a "baseline" population of about 250 million people. (Some sweeping catastrophic events, such as widespread plagues have effectively caused the population to fall by half at some times in the past.) He goes on to explain that from 1 AD to about 1650 AD, the population doubled to 500 million (0.04% annual growth rate); it doubled again to one billion by 1850 (0.35% annual growth rate); again to two billion by 1930 (1% annual growth rate); and again to four billion by 1976 (1.5% annual growth rate).

The world currently holds about seven billion people and would add an extra one billion by 2023 if it were to continue on the current rate of growth. Doubling, at this high 1.5% annual growth rate, becomes absurd rather quickly. Indeed, in just 200 additional years, the world's population will double four more times and sit at about 130 billion, which will require roughly 8,000 people to live on every available square mile of arable land—an unlikely (and unpleasant) scenario. (That also equates to every person having 18 feet by 18 feet of living space—and nowhere left to grow crops, let alone hike or hunt or sit and watch the sunset!) No matter how large a thing you are dealing with, exponential growth ultimately becomes unrealistic and, more importantly, unsustainable. This is a key point that defines how energy will be used in this century.

5.2.2 PHYSICAL LIMITS

As was discussed in Chapter 1, both nonrenewable and renewable energy are governed by limits. Thus, it is necessary to acknowledge that one of the primary facets of sustainability is one of limits—either imposed by the natural world, or imposed by humanity. Moreover, a stock-based energy resource must, by definition, have finite stocks: constrained by the finitude of the system within which they exist, while a flux-based energy resource will be limited by its natural, renewal rate.

In this era, the stock-based resource limit of greatest concern is petroleum. Petroleum, or more commonly oil, has been called by some the "life blood" of the world's economy. It drives practically all of transportation, and global demand has been generally rising for a century and a half, only curbed by significant economic factors. But because oil is ultimately finite, sooner or later, the supply must be exhausted. It would seem that there can be details to argue about in terms of *when* to expect this to happen, or even about what oil shortages will be like, but there should be no argument that exhaustion will happen. Yet, because humanity has been able to discover more and more and at a pace that seems to match exponentially growing demand, this issue of reaching a truly physically limited peak in oil production—or "peak oil"—has seemed to garner a "never cry wolf" sentiment amongst many that are comfortable with current consumption patterns. But, one should be assured that there are limits to petroleum production, and those limits are dictated by the finitude of oil (and gas) in Earth's crust. We will discuss this in detail in the following sections.

5.3 PEAK OIL: UNDERSTANDING OIL LIMITS

Petroleum production has been growing exponentially virtually since the industry began in 1859, in response to growth in global demand for energy-rich fluids. Of course, nothing in the universe can grow forever, especially at exponential rates. Sooner or later, global petroleum production must reach a peak and begin an inexorable decline to ultimate exhaustion. We probably cannot know until we are a few years past when the peak will come (or rather, came), what its decline will be like, or how much oil will eventually be drawn from Earth's crust. But, the peak will create considerable pressure on an energy transition, so we must seek to understand as much as we can.

The character and the timing of a global peak in oil production are determined by consumption, economics, and technology—three conditions that are always changing. Thus, assertive predictions of peak production are nearly untenable because it requires the predictor to have an excellent grasp of these three determinants, even going into the future. Yet, the predictor can be in the right "ballpark," tempering the sometimes wild optimism that seems to prevail in this era of continuous growth. Indeed, some optimists run afoul even of the physical limits of the known universe, let alone of our Earth. Of course, far short of the limits of the universe, oil and gas extraction are defined by Earthly limits—Earth itself is definitely finite. But oil and gas formations occur within sedimentary basins, and there is an even more limited amount of sedimentary rock on the planet. Moreover, oil and gas cannot occupy all the sedimentary rocks either. These constitute very real limits.

5.3.1 SPECIFIC DETAILS

Let us consider a few specifics of the peak oil concept. Just as it is absurd to think of an indefinite exponential growth of the human population, it is also absurd to expect continued exponential growth in the consumption of a finite resource. However, the simplest view of petroleum limits is that of exponential growth, shown in Figure 5.1 below. In this scenario, production growth continues until all of the resources are consumed and then production plummets instantly to zero.

With knowledge of the mathematical function for exponential growth, we know that, indeed, it cannot practically happen this way. If it did, we would be in the incorrect "running out of oil" scenario many people hold about the depletion of finite resources. In this scenario, an exponential growth in production would proceed with time until the moment before the resource stock fully depletes. This translates to oil production at full capacity one moment and the next moment that capacity has fallen to zero. Not likely.

Instead, we have come to imagine consumption/production to proceed in a more realistic, but still extremely simplistic fashion. In Figure 5.2, we depict the following scenario:

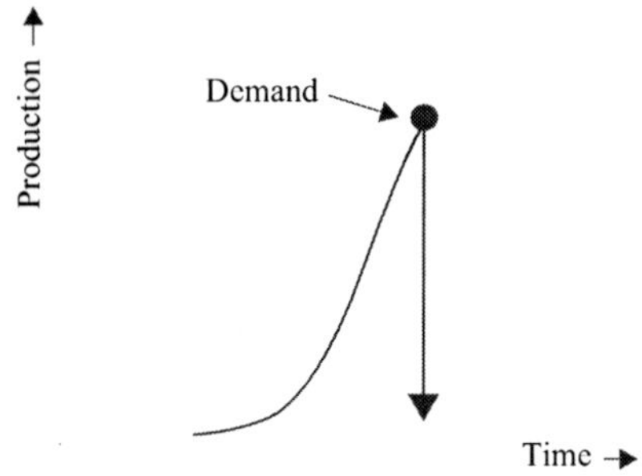

Figure 5.1. Exponential growth and collapse, the common notion of suddenly "running out of oil."

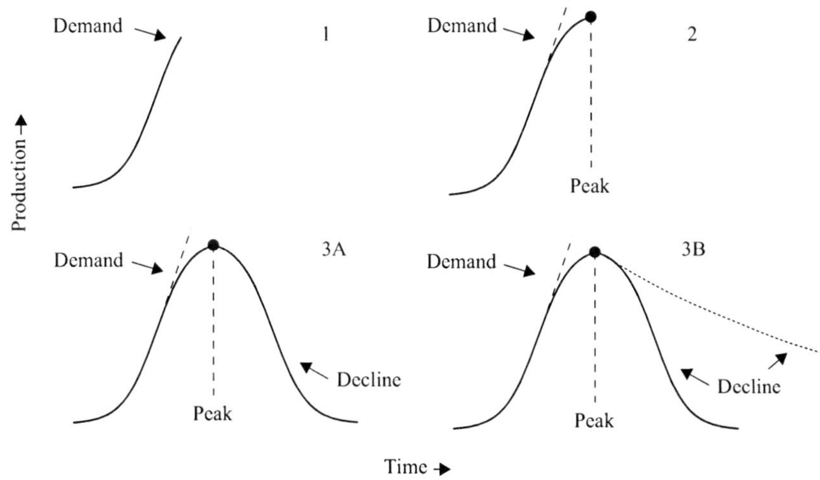

Figure 5.2. A simple oil consumption/production cycle.
Source: Based on Ebenhack and Martínez in Khodeli (2009).

1. Consumption/production will begin as an exponential growth function.
2. As new sources become harder to find, prices will rise and fuel substitutions (either alternative energy or conservation) will take over some of the market. This trend will begin in earnest, when the inflection point is reached, beyond which growth cannot occur at the same rate and will continue until consumption rates peak. As long as growth in production capacity can keep pace with demand, there is little upward pressure on prices.
3. Finally, a declining production trend will follow as high prices and resource substitutions prevail. This decline may be precipitous (3A) or gradual (3B).

This simple construct depicted in Figure 5.2 is at the heart of what is known as "peak oil." This refers to the peak in global oil production, whereby roughly half of the original reserves have been consumed, and oil production will not only stop growing, but begin a terminal decline. That halfway point represents the peak. M. King Hubbert is attributed with starting the use of the Gaussian, or, normal distribution function, intuiting this more plausible production sequence and received a tremendous amount of attention when he was fortunate enough to predict the approximate year that oil production would peak in the contiguous 48 United States. We note that Hubbert was extremely conservative, and wildly incorrect, in his estimate of the total production output number, but an accurate prediction of the date assured continued attention for his work. Yet, as with most things, the devil is in the details, and the details are extremely important once you start trying to use this mathematical intuition to then extrapolate and predict the timeframe for the total demise of the energy source that represents the backbone of the world's entire transportation sector. A few of these details that should inform peak studies are expressed below.

5.3.1.1 Detail One: A Stacked Deck

The first detail that should be clarified is one that Maugeri raises in *Age of Oil* (Maugeri 2006). Hubbert had the proverbial deck stacked for him to make a very good estimate of oil's "demise" in the contiguous 48 states, because it is arguably the most intensively explored province on Earth, having been prospected since the 1800s. While oil production did not start in the United States (nor was Drake's famous well the first source of petroleum production in the United States), it was seminal and quickly made the United States the hub of petroleum exploration and production. A new industry sprang up in response to a new understanding of petroleum's potential and how to obtain it in unprecedented quantities. Again, according to Maugeri, there have been roughly two million wells drilled (one million alone in Texas) and about half a million currently producing wells in the United States, making the subsurface very well known to geologists and amenable to estimating the total resource available in the region. Most other regions in the world can claim to have currently producing wells on the order of 1,000 and exploration not nearly as extensively completed as in the United States.

5.3.1.2 Detail Two: Demand Curve

The next detail that should be clarified is that when a physical peak is reached, the demand curve will tend to remain exponential (depicted in Figure 5.2). This is known by economists as a market failure. Indeed, the rate of consumption in the United States, while reduced from the pre-1970 peak, continues to grow at an exponential rate. The market failed to keep pace with demand locally, so new provinces were explored elsewhere and developed to help maintain this increasing demand for oil. We must not forget the value that cheap, abundant, versatile, and portable energy provided by oil offers to economic development. The affluent world will have some tendency to try to keep expanding this benefit, looking for new and unconventional sources to meet demand. Moreover, the Developing World, led by China and India, will cause continued upward pressure on global demand for petroleum energy.

5.3.1.3 Detail Three: Peak Character

The next detail involves the character of the global peak, itself. Will it really look like a peak or will it plateau for an extended period of time? And will there be multiple peaks, representing bimodal or log-normal distributions? Finally, there is the character of the downward trend. Most peak oil commentary assumes that the decline will be precipitous due to the notion that the energy needed to extract and refine exceeds the energy derived from the resource itself. Petroleum, though, continues to have an exceptionally high "energy profit" associated with its production, compared to many competing resource. The precipitous decline also discounts the fact that a great deal of oil has been discovered and left unproduced because it was "sub-commercial." As shortages appear and prices rise, those abandoned resources will become economically attractive. These undeveloped prospects will not have lower "energy profits"—it was economic factors based on an export market focus that caused many of these wells to be condemned, not the amount of energy required to extract them.[3]

Furthermore, while some of the enhanced oil recovery (EOR) methods discussed in Chapter 2 do have lower energy profits than primary or secondary production, the energy

[3] Co-author Ebenhack was charged with making those recommendations on international exploration for a major oil company in the 1980s and personally condemned a number of wells that would have been very profitable if the intended markets had been local.

gained would still be quite large compared to many alternatives, so EOR techniques will almost certainly be employed to help slow the global decline in petroleum production. Remember that twice as much oil remains in the reservoir as has been recovered after a field has been completely depleted, so there is a great deal of opportunity to tap this unrecovered oil.

Dr. William McNeely, a physicist retired from Boeing Corporation, offered an interesting perspective on this problem in a public lecture at Marietta College (Ohio) in 2011, noting that people have often worried about running out of oil and that, so far, more has always been found. He observed that drilling to find new reserves costs money and companies do not spend money to satisfy curiosity. The drilling and new discoveries have proceeded at a pace that has met demand. Of course, the point at which a shortage arises is the point at which new drilling becomes less successful at replacing reserves. It will be hard to predict that point *a priori.*

5.3.1.4 Detail Four: Descriptive Versus Prescriptive Analysis

The final detail presented here comes from Hubbert, himself. According to Maugeri (2006) and Deming (2000), Hubbert explicitly says that the production cycle estimates he made initially were not derived from a mathematical function, but simply "tailored by hand." It is unclear when Hubbert's work changed from being descriptive to becoming prescriptive for forecasting oil's ultimate demise as a primary resource. Probably, it occurred when he began to receive substantial media attention during the energy crisis of the 1970s and then again as we got closer to his global prediction in the late 1990s. From his earlier papers, it appears to us that he did not view his work as being inerrant in the way that his modern acolytes seem to. He raised an alarm that petroleum production in the United States must inevitably peak and followed that with some predictions for the United States and then for the world at large. It is worth noting that he revisited and revised his own predictions.

5.3.2 ANALYSIS

Despite the details listed above, which are problematic for those that use the model prescriptively and definitively, Hubbert's work was indeed invaluable to the need for American society to temper their consumption habits with the fact that exponential growth cannot persist beyond short periods of time. Hubbert's legacy is an important and lasting one with regard to the significance of his findings on depletion and its influence on understanding and public debate on resource futures in the 1970s and today. Indeed, although Hubbert transformed previous random speculation of oil's demise into a more concrete line of inquiry using reasoned methodology, the public found no particular interest in it when he first published his findings about a future peak (Hubbert 1949), but was especially interested when a prevailing energy shortage in the United States occurred in the 1970s. The accuracy with which he predicted the timing of the United States peak more than 30 years earlier was striking. It was then that the cultural significance of his work became evident and long lasting.

Interpreting the scientific activities that have come to shape peak oil should better inform the conclusions that were reached by Hubbert and others. For example, the current peak oil line of inquiry largely comes from the geologic sciences (Campbell 2002; Deffeyes 2010), which represent the technical understanding that undergirds the reality of absolute limits (i.e., the finitude of geologic basins that can contain hydrocarbon accumulations and the processes by which biotic material is transformed into oil or gas.) The geologic sciences also shed considerable light on exploration activities and discovery rates. They do not, however, represent the knowledge base of extraction technologies or operational considerations. Engineers bring this understanding and are peculiarly under-represented in peak oil analyses. This is an especially important point because Hubbert appeared to say little about the character of resource depletion after the production peak, a point where extraction technologies and operational considerations become important (See Ebenhack and Martínez in Khodeli 2009).

Because of this, the argument over oil limits continues in popular discourse to be dominated by extremist views that either claim that the shortage is upon us already, with a steep decline imminent—or that there can never be a shortage at all. On the cornucopian side, there are those who assert that we never need to worry about petroleum shortages. Their optimism is largely based on an absolute, and in our opinion, misplaced, faith in the economy. A (relatively) free market can be highly effective in allocating resources, but it remains dependent on the availability of real resources to allocate (Ayres 2007). Whereas it may well be true that, given time, the market will provide money to support research, development, and deployment of new energy technologies, the reality is that the marketplace has not yet provided significant support to the development of alternatives. Indeed, so long as petroleum and related products are able to supply energy cheaply and abundantly, the marketplace really provides a counter-incentive to alternatives.

We cannot realistically expect support from the marketplace for alternatives until shortages are prevalent. Then, it may be too late to support a smooth transition. This argument is based, in part, on the limits imposed by time. Whereas there is clearly truth in the potential for the marketplace to allocate investments, research, and development to find alternatives, it is never an instantaneous process. We believe that there is ample evidence to suggest that there are impending shortages and precious little time for the markets to respond. Fundamentally, the marketplace is at its best in conditions of abundance of raw materials, it is of questionable value in a crisis.

Consider the Titanic. There was a great deal of wealth on board. There was a great deal of demand for lifeboats. There was a demand, a willingness to pay, and an ability to pay. Why weren't lifeboats constructed when the great ship began to sink? It wasn't impossible—there were many materials on the Titanic that could have been used to make boats or rafts. Time was the problem. The boats did not preexist and there was no time for a marketplace to create a product to meet the demand. Certainly, this is not a perfect example of what can be expected when a petroleum shortage arises. Several flaws can be identified in the analogy. But we do think it helps to illustrate that the marketplace needs time to respond and is limited in what it can do in a crisis.

What also hampers the cornucopian argument is optimistic assertions have gone so far as to claim that "there is enough of everything to meet the needs of a growing human population for seven billion years" (Simon 1996). Seven billion years! That sounds very reassuring, but it's an incredibly long time! The claim suggests that there are no plausible limits to any resource. We can use this assertion as an example to help us see that all physical things have limits: even the universe itself.

Let's return to the mathematical understanding of growth and consider, as was done in Ebenhack and Wilson (2005), a mere 1% per year population growth (which is on the low side of current population growth rates). In seven billion years, that would result in a human population of $10^{30,000,000}$ people. That number is 10 followed by 30 million zeroes! An old college roommate, Richard Talcott, is now editor of Astronomy magazine, so coauthor Ebenhack asked him for an approximate estimate of the total mass of the universe. He replied that studies suggest a known mass of about 10^{53} pounds. Because there is matter that we don't fully understand yet, he suggested multiplying by a factor of 100 to account for that—resulting in a total of 10^{55} pounds, including all the stars, planets, black holes and things we don't understand. Alluding to the human population in seven billion years calculation, he added, "feel free to check my math, but I doubt that I'm wrong by more than a million orders of magnitude." There's nowhere nearly enough mass in the universe to continue making more people. In fact, at 1% per year population growth, it would only take about 10,000 years for human bodies to require all the mass in the universe. Unlimited growth is impossible. The cornucopian optimists are wrong. Humans must understand and work within limits (Ebenhack and Wilson 2005).

Nevertheless, the pessimists often paint an unnecessarily bleak picture. There is a common tendency, when trying to raise an alarm to make that alarm shrill. There may be merit in "getting people's attention." James Howard Kunstler wrote the colorful and compelling book, which we do highly recommend, *The Long Emergency* (Kunstler 2005). Quoting him in an article, a few years ago, one of us felt the need to describe him to readers who may have been unfamiliar with his work. The phrase "deliberate provocateur" came to mind (along with a note that Kunstler's work was, nevertheless profound and thoughtful). An email was sent off to him, with the paragraph so characterizing his work and saying, "if you're offended, I'll change the description." Kunstler replied promptly, "It sounds about right to me, good luck with the article." Provocation can be good and valuable to move us from the inertia of the status quo. Yet, provocation must be tempered with facts and reason. We wish to bring you the realistic perspective that can help to sort through the confusing claims on an important topic.

Confusion is understandable. There is a great deal that we cannot know about the limits to petroleum or the transition to alternative energy systems. Nevertheless, there are some things we can know—and a great many things that we can know will not be true. Deduction was the basis of Sherlock Holmes' mythical accomplishments, "After eliminating the impossible, whatever you are left with, no matter how improbable, must be true." We can make some use of this approach in real life to operate in uncertainty. Due to the deductive reasoning approach, we will also employ "*reductio ad absurdum*" analytical methods to help determine those things that we can know are not true.

There are a few realities about which we can be clear: petroleum production cannot continue to meet an ever-growing demand. Even if it could, there are other problems associated with its use that should motivate us to seek alternatives; nevertheless, the fossil fuels have provided a rich basis for growth and development for half of humanity. The nature of a "peak" tells us that the world will be producing more oil than ever before at the peak—and near record levels shortly after the peak. It will not be a sudden disappearance of petroleum. Finally, as we will continue to demonstrate, the challenge is that, even as the affluent half of the world must strive to conserve its resources and find alternatives, the least affluent half needs more access to energy and other resources in order to achieve reasonable development.

There is a common alarmist perception that global petroleum production already has peaked or will peak very soon. The latter may be correct. The former is almost certainly false. If demand accelerates (perhaps in response to continued economic growth in the world's two largest population centers, China and India) the world's surplus production capacity may be exceeded relatively soon. The extreme pessimists, though often cite the year 2005 as the point at which the world passed peak petroleum production. This notion certainly enhances the sense of urgency, but it seems to miss both the physical realities of petroleum production and the history of petroleum predictions. Many such predictions have been advanced in the past and always proven to be wrong, so far. Do we now have clear evidence that petroleum production has already peaked? In fact, global petroleum production has continued to increase in the years after 2005. There was a dip around 2007 and 2008, but that was most likely a response to high prices and the onset of a global recession, both of which tended to suppress demand. Since 2010, with the consequences of the recession seemingly easing, it appears that petroleum demand and production have both been on the rise again (WWI 2010).

Predictions of its demise have plagued the petroleum industry since its birth. Repeatedly, doomsayers have suggested that the world had no more than 20 or 30 years of oil left (Deming 2003). The repeated failures of these dire predictions have led some to dismiss the finitude of petroleum resources completely—even claim that the world can never run out of oil. We suggest that the "boy who cried wolf" is an excellent metaphor for predictions of global petroleum shortages (Akins 1973; Maugeri 2004). There have been many predictions that history has shown to be ludicrous. The fable of the boy who cried wolf, though, is not a proof that there is no such thing as a wolf and that one should never worry about wolves. Indeed, it is a story of precisely how disaster befalls the people who cease to pay attention to any claims of seeing a wolf, which finally comes and devours the flock. (We think, in the original version, the wolf ate the little boy too and we are inclined to say, "good, he had it coming.") The fact that previous claims about running out of oil have been proven wrong means nothing about current or future claims. Still, one should be very careful about crying wolf. The optimists fail to understand the fable by sitting back, saying that current predictions must be wrong because they've been wrong before, while the pessimists fall into the trap of being little boys crying wolf, by trying to exaggerate their claims without adequate basis. To avoid the trap, we assess the impending global peak in petroleum production by making deliberately optimistic assumptions as we assess the gloomy, but inevitable, reality of shortages. All peak calculations start with an estimate of how much petroleum may ultimately be recovered from Earth: that equates to the area under the curve of production rate vs. time. While most peak oil theorists currently active use approximately two trillion barrels for that ultimate potential recovery figure (see Campbell 2002), we would rather choose six trillion barrels. It is the highest estimate we've found published and three times more optimistic than the peak oil theorists (Grossling 1977). So, a reader should expect that we will be more likely to err on the side of optimism than pessimism. We think it's valuable to proceed with deliberate optimism for this argument. It obviates arguments about "what about the new Utica and Bakken Shale plays, or the new discoveries in Uganda, or EOR, or …?" We think that there will be lots of oil (and gas) reserve additions, but there is still a finite potential. We choose to use a high-end assumption about potential to avoid crying wolf prematurely. (Though a six trillion barrel ultimate recovery still only pushes a peak out by a few decades, if high consumption rates persisted.)

What signs will warn us that the peak is near? This is a highly problematic question. We do not believe that there is any realistic marketplace phenomenon that can be expected to predict a shortage in petroleum. As long as it is available in sufficient quantities, the marketplace will merrily allocate resources. Surely, there are many analysts trying to predict future oil supplies—many of them looking at political factors that might affect short-term oil supply. When we get into a global petroleum shortage, we can know that oil prices will rise (and other energy prices are very likely to follow.)

Indeed, oil prices reached $147 per barrel in 2008. So, this seemed to be proof that the pessimists were right, that global petroleum production was already reaching its peak and demand was exceeding supply. It was, however, apparently an aberrant and exceedingly short-lived price spike. Since the price collapsed by more than 50% within a few months, this can hardly be seen as evidence of having reached the peak. The spike seen in 2008 was in part a function of the declining value of the dollar (it took more dollars to buy anything internationally) and of speculation in response to claims that the peak had already occurred in 2005. But the price collapsed when the dollar firmed up and the speculation dried up.

In fact, petroleum demand shrank in response to the high prices, especially in the United States, where a good deal of consumption is discretionary. Dr. Don Stinson, petroleum consultant and former long-time chairman of the Petroleum Engineering Department at the University of Wyoming observed, "The United States doesn't produce enough oil anymore to affect the supply side of the equation, but we sure can affect the demand side!" Prices respond to supply and demand—or even to perceptions of the balance between supply and demand. Any passing shortage in the marketplace raises the specter of "running out of oil" and drives prices up dramatically, which tends to suppress demand quite quickly and to increase supply through increased production and exploration efforts, with much longer lead time.

5.3.3 A CLOSER LOOK AT THE CHARACTER OF A PEAK

There is an interesting paradox. The optimists who claim we will never run out of oil are correct in an absolute sense, while being ridiculously wrong in the practical sense. It is true that no matter how high the price of petroleum becomes, ultimately, no technology will ever extract every drop from the proven, producing reserves—let alone find every small reservoir that exists. The decline is neither total nor sudden. There will always be some oil left in the ground. Even in the highly flawed and simplistic running out of oil analogy, there are almost always a few drops left. Furthermore, new oil is being created: perhaps a few thousand barrels a year. In absolute terms, the world will never completely run out of oil. Shortages will cause price to escalate, which will create the incentive to find and recover more of that remaining oil.

But in a practical sense, a crisis will come when production rates can no longer keep pace with demand growth. Consider that at the onset of the shortage, we will be producing more oil than ever before! That is the nature of a peak. As Hubbert noted a half-century ago, petroleum reservoirs, provinces—and ultimately the world—will undergo production growth to a peak or plateau and then enter a long, gradual decline in production.

The pessimists are right to be warning us that a potential crisis is imminent. We believe that exaggerating how soon the peak will come (e.g. "it already happened") and how dramatic the decline will be are their only real errors. The world needs to be warned. Even with good information, it will be hard to predict the peak accurately before it arrives. Since much of the information on reserves and even production capacity is hidden under the clouds of corporate and national secrecy, with varying incentives to be optimistic or pessimistic on the parts of all the disparate players, we do not believe that we can expect to see any clear warning signs until shortages arrive.

Therefore, we cannot know exactly when worldwide petroleum production will begin to decline, let alone the details of how the transition from exponential growth to exponential decline will unfold, until the decline is underway. Indeed, we will probably not be able to identify the global production peak until some years after it has occurred. However, remember some of the things we can know. Petroleum production will reach a peak and will ultimately decline. That peak will occur when global *surplus* production capacity has diminished to zero and demand continues to grow (unless some alternative source succeeds in carving out a major market share very soon). If the character of the decline is a function of global production rates, it will be a long and probably very shallow decline (Hirsch 2005).

Returning to the original United States peak production work of King Hubbert himself, he did not predict timing, but rather focused on the inevitability of the peak phenomenon. Hubbert drew a dashed line from the peak, which was relatively flat. It appears that, only when his work drew some public interest and questions about when the peak would occur, did he begin to use a symmetrical (bell-shaped) curve (Ruedisili and Firebaugh 1975). This simplification enabled him to make a plausible estimate of when the peak would occur, based only on information about the rate of growth in demand and the total amount of ultimately recoverable oil (which would be the area under the curve of barrels per time versus time.) Without making some assumptions, it is impossible to predict future events. Hubbert did predict with surprising accuracy that the peak in United States production would occur in the early 1970s. While he was right about the timing, he quite substantially underestimated the height of the peak. The United States actually produced much more oil than he guessed, but our demand for oil also increased faster than he expected, so those two factors offset each other, resulting in an accurate estimate of the timing of the peak.

Later followers of Hubbert, though, have made the shape of the curve a central part of their "Hubbertian" dogma. Are they right? We can certainly not think of any physical reason to expect symmetry. The growth side of the curve is governed by demand, as long as the resource is sufficiently abundant so as not to provide any constraint. Once the physical limits of the resource begin to constrain the rate at which production can grow, an inflection point can be seen—the curve continues to climb, but begins to level out. This happened for United States production, but it didn't cause any immediate shortage because there was plenty of oil available to import from other parts of the world. That, of course, won't be the case when we reach a global peak—there will still be new discoveries, with new places from which to import oil, but the definition of the peak is that those new discoveries will no longer be added fast enough to maintain the previous growth rate.

Note in Figure 5.3 how clearly the "rounding off" can be seen on the curve of United States domestic oil production as it approaches the regional peak (log-linear scale). There is a brief decrease in the curve, which appears to correlate to the Great Depression. After that, the growth resumes, but the rate of growth is clearly diminishing from about 1930 until the actual peak around 1970. As soon as we reach that inflection point and oil production can no longer keep pace with exponential growth in demand, an entirely new set of dynamics will come into play.

A seldom-discussed point is that the actual shortage should be expected to precede the peak by several years, because demand is tending to increase exponentially and the rate of growth of petroleum production will begin to "tail off" toward the peak. At that point, several years before the actual peak, the gap between extrapolated demand and real production will constitute a shortage. Like a real hill, the curve starts flattening out before reaching the actual peak. The point at which the curve departs from exponential growth and begins to flatten out (the inflection point), growth continues, but the *rate* of growth is slowing. There will be a gap between the amount of energy the world's people seek and actual production. Demand will exceed supply.

Is that possible? Economic theory is based on supply and demand balancing each other to establish "market clearing" prices. Indeed, the principles of "neo-classical" capitalist economics

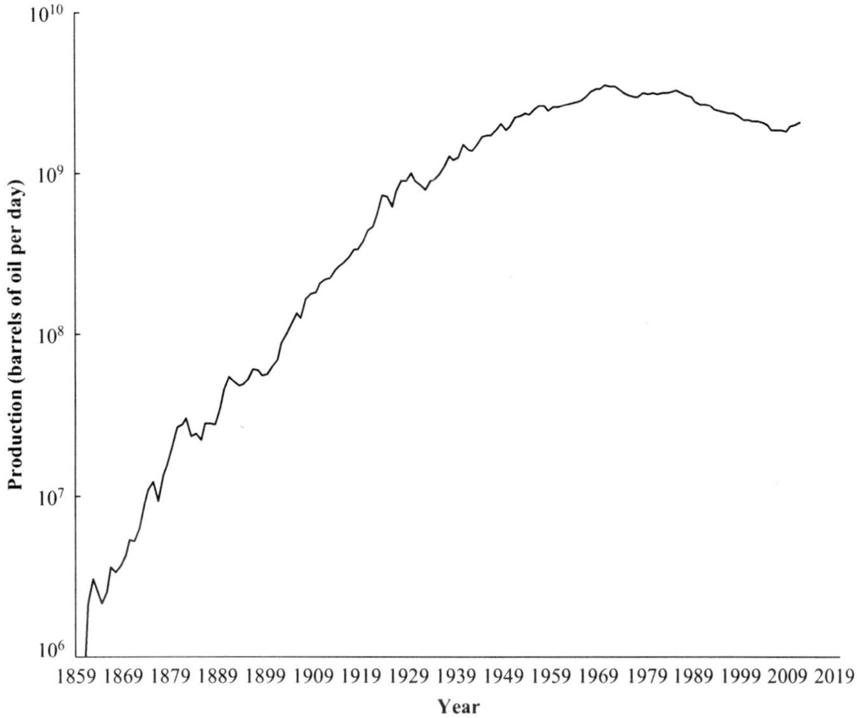

Figure 5.3. Oil production data in the United States, 1859–2011 (log-linear scale). Note the clear "rounding over" or inflection.
Source: eia.gov

are built on several assumptions, including unlimited substitutability (Ayres 2007; Ruth 2008). That is, anything in short supply can be replaced by something else. We suggest that essential materials, such as air and water do not have ready substitutes. Energy also serves critical human needs, so while a new energy source can substitute for another, it must be able to replace the quantity and quality of the previous energy source.

Certainly, there is reason to believe that the peak may not resemble a "bell curve." Wolfgang Schollnberger (2006) has discussed a variety of shapes that may describe the peak. For many of the reasons just discussed, he observes that the shape is not predetermined. Figure 5.4, based on his work, shows a general range of characters possible. He points out that market forces will still influence production rates into and after the peak time, because higher prices will undoubtedly promote more exploration, reserve additions in existing fields (oil that was not economically recoverable under lower prices), and conservation measures. We think that the most likely character of the peak oil curve will be something between a long plateau and a slow, shallow decline. However, only hindsight will tell us the actual timing and character of the peak.

5.3.4 *WHAT WE CAN KNOW*

Whereas we cannot expect to make precise predictions of the timing of the peak, let us consider what we can really know about the peak. As we said, it is inevitable. Everything has its limits.

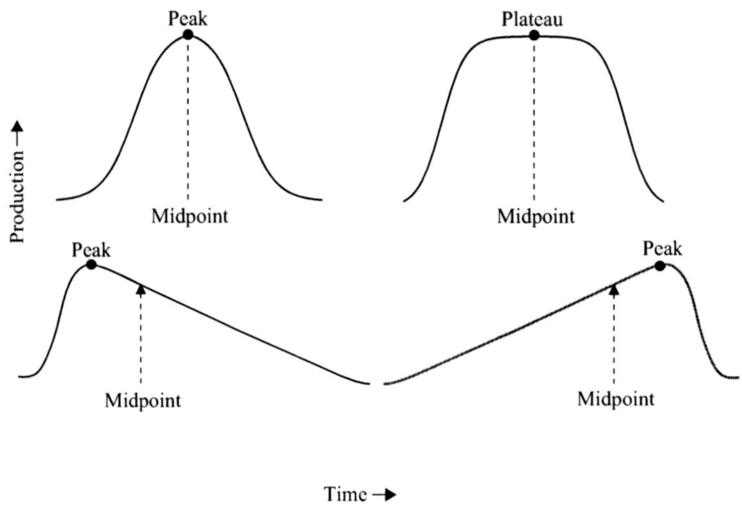

Figure 5.4. Various peak oil production-based scenarios.
Source: Schollnberger (2006).

A peak of some sort is inevitable as long as we consume oil at a rate that exceeds Earth's rate of making new petroleum. Currently we estimate that consumption exceeds the rate of new petroleum formation in Earth's crust by a factor of about a million. That cannot possibly be sustained indefinitely. What is even less sustainable is that the rate of consumption has been growing exponentially for 150 years. We have discussed what a powerful phenomenon exponential growth is. Consider that in 1860, the world's petroleum production (essentially all from Pennsylvania), had quickly grown from nothing to a few thousand barrels per day—150 years later, it has exceeded 87 **million** barrels per day, when unconventional oil is included[4] (WWI 2010).

Petroleum will fail to meet growing demand at some point. It is rather stunning that petroleum has been able to maintain that growth rate for so long—which explains why so many people "exaggerated the rumors of petroleum's death" over the years. But just as the exaggerated rumors of Mark Twain's death eventually came true, so will predictions of petroleum shortages.

The "sharp decline" peak analyses seem flawed in two respects. First, we really don't—and can't—know how much oil exists in Earth or will ultimately be recovered. The analyses do not adequately account for the effects of price changes and thus, the nonlinear nature of the decline. Detailed analyses of the American transition clearly demonstrate production rate trends that are strongly correlated to price. As depicted in Figure 5.5, the years prior to the 1973 Iranian Embargo, when prices were characterized by decreasing constant value, the peak and the decline were indeed sharp—nominally symmetrical with growth. However, approximately four years later, an inflection can be seen in the decline data, indicative of a slowing decline. In 1979, the second price spike (following the Iranian Embargo) was superimposed on the decline curve, which actually resulted in a reversal of the decline trend. By 1981, prices had collapsed, yielding a return to a steeper decline rate approximately four years later.

Analyses of the data reveal several things. First, there is a clear positive production response to increased petroleum prices and a related negative response to falling prices. This

[4] Many pessimists do not include unconventional oil in their peak analyses, which seems curious to us. An unconventional method almost always evolves into a conventional method, given enough time.

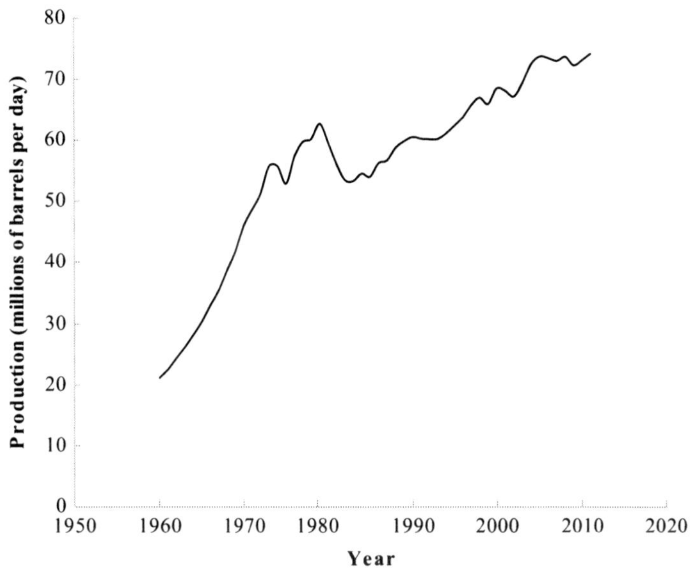

Figure 5.5. World oil production from 1960 to 2012.
Source: eia.gov

is not terribly surprising in terms of oil field operations. Rising prices increase the incentive for more intensive exploration, more aggressive completions, and greater application of EOR technologies. All of these responses tend to increase production (and reserves). But all of them require time to mature. Both positive and negative responses to rising and falling prices tend to lag behind the price change by at least three to five years. Increasing both drilling activity and enhanced recovery technologies takes time, often up to 10 years for a variety of reasons stemming from the time it takes for corporate planning to respond to higher prices, followed by the time it takes to implement new projects and see the first new production being realized from them. And this is just the first response from the efforts with the least time—some of it simply relating to returning marginal wells to production.

Lag time is one of the issues that lead authors without petroleum operations experience sometimes tend to denigrate the potential reserve additions from EOR and increased drilling activity. Strong oil prices have never prevailed for more than 10 years at a time. Thus, projects with long lead times have just reached the market at the time when prices began to fall—creating an impression that the technologies have failed. Furthermore, drilling and discovery trends are not simple functions of the amount of oil remaining to be discovered. It is true, of course, that many of the largest, most readily accessed fields have been discovered and placed on production decades hence. As time goes by, there will be less big fields to discover. Statistically speaking, we are sampling without replacement. It is like playing blackjack rather than roulette. Each ace dealt is one less ace remaining in the deck or decks.

Nevertheless, we are not at a saturation point, in which every geologic province has been extensively explored and developed. The United States has millions of wells drilled, while the entire African continent, with six times the areal extent of sedimentary basins, has only a few thousand wells. Some authors might reply, "but that's because there is no significant oil being found there." They would be missing at least two important points. First, exploration is still a high-risk enterprise. The geophysical advances in describing the shapes of subsurface features have done relatively little in terms of increasing our ability to tell which formations hold oil or even gas. Therefore, even big fields are not necessarily found with the first wells drilled in the area. Second, the evaluation criteria for international exploration are radically different from local exploration projects—for economic reasons. Coauthor Ebenhack has personally been involved in condemning multibillion dollar discoveries as subcommercial, because the costs of developing them for the export market were in the same price range. This assessment is very much a function of prices. The higher the oil or gas price, the more projects move from "sub-commercial" or "marginal" into "commercial" categories. Indeed, increased prices will stimulate reevaluation of many prospects that were abandoned as subcommercial. There will be a great deal of oil and gas "rediscovered" when prices rise and stay higher.

Furthermore, recall the discussion of the character of oil and gas reservoirs. They are complex rock structures, containing the oil and gas in myriads of tiny, interconnected pores—along with water, as explained in greater detail in Chapter 2. The recovery mechanisms are complex. Some technologies, such as fracturing, can greatly increase the production *rate*, but not indefinitely. Indeed, stimulated wells often show very steep initial declines, as oil or gas flow from the contacted rock into the induced fractures, but then the fluids must flow through the low permeability rock itself, which is much slower. The decline levels off to a very low percentage after a while and wells may continue to produce for decades at much lower rates. So, some projects can add short bursts of impressive new production, followed by long lives at low production rates. This will almost certainly be true of the shale plays being developed in the early 21st century. The shale reservoirs have radically changed the reserves picture in the United States and are likely to transform reserves around the world. The demonstration of substantial production and reserves potential in the hitherto ignored shales certainly makes clear that the global petroleum production peak did not occur in 2005, as some would have us believe. We are continuing to expand the kinds of reservoirs from which we can produce oil. This will extend the peak and the area under the curve for a while longer. New production can mitigate regional or global decline trends. Even in the stabilized price region of the 1990s, United States decline rates were leveling to something at or below 1% per year again. The marketplace will support more petroleum development, but the added production does not come instantly and often not at extremely high flow rates.

While we have been rapidly drawing down the global petroleum resource base, the industry has been able to add new reserves to maintain a production capacity surplus. The world has always been able to produce at least a little more supply than the marketplace has created demand. The crisis, or systemic global shortage, can be described as the point beyond which new discoveries and reserve additions will no longer be able to support a production capacity surplus. A surplus is essential to normal behavior of petroleum markets. Like air, or water, or food, energy must be readily available to support human life and enterprise. It is an essential human need and must be available in excess. When there is barely enough to meet demand, prices will rise in response to the shortage, and some demand will diminish in response to higher prices. More production both of petroleum and of alternatives will be incentivized, but neither will be able to respond instantly.

The more prompt response in the affluent world will be conservation. This near-term conservation will consist of foregoing some discretionary consumption. It is easy to elect not to make a trip to Disneyland or cross-country to visit grandparents. This belt-tightening is seen in every petroleum price spike, but the belt is loosened up again and the deferred consumption activities are taken up. Conservation measures that are, themselves, sustainable take longer, such as living closer to work and in smaller houses. Over time, we will find ways to do more with less, but the shortage time will be disruptive.

When can we expect this disruption? Although we cannot know how much oil may be ultimately recovered from Earth's crust, or when the peak will occur, or exactly how the peak and decline will unfold, it is not important for us to know any of these things precisely. Just as it is not important to know when or how we will die. We can still live our lives understanding our mortality—and we can know reasonable limits. We can also understand the practical finitude of petroleum and develop reasonable expectations as to when the peak must peak and how the decline is likely to proceed. The peak is likely to occur soon enough (and almost certainly by the middle of this century) that we should take action to prepare and mitigate its effects. We can be certain that prices will rise rapidly when the world reaches a global peak in petroleum production. We can have confidence in rising prices, because we can readily see that demand cannot sustainably decrease worldwide and that time will be required for alternatives to take up significant new market share. The needs of the Developing World will present urgency for some continued growth in energy production and consumption. For the affluent world, conservation will be one of the few options that can respond in the timeframe required.

5.4 LIMITS OF OTHER RESOURCES

While most of this chapter has focused on oil limits, it should be obvious that what is described here also applies to natural gas, coal, uranium, and in many instances biomass, as they all, like oil, represent reserves-based, stock-limited energy. (Remember, biomass is used as a reserves-based resource, and although biomass's renewal rates are faster than fossil fuel renewal rates, if consumption outpaces renewal and harvesting destroys the local ecosystem, biomass becomes nonrenewable energy, often very quickly.) If these resources are used to near-exhaustion, all of these stocks will decline in ways similar to oil. Of course, each resource will have different decline characteristics, impacted by the total reserves available and by consumption behaviors, as described by Schollnberger (2006). This is important because when authors refer to coal having sufficient reserves to last 200 years, what they really mean is "at current production levels." If coal takes over more of the global energy market, then the 200-year number shrinks. It shrinks even more dramatically if exponential growth in demand is factored in as well. The same can be said for nuclear fission, oil sands, etc. The longevity of a finite resource is very much a function of its rate of consumption.

It is also true that the flux-limited resources, which cannot be depleted by human use, likewise face limits. Recall from Chapter 1 that the term "flux-limited" reflects the reality that there is a finite amount of energy that we can tap from these resources, usually tied to the rate at which we receive that raw energy. Furthermore, we cannot tap them at a large scale without disrupting other facets of Earth systems of which they are a part (including land, mineral resources, and water). However, if we consider flux-limited energy such as hydro energy, wind energy, solar energy, and geothermal energy, we should realize that the peak production cycle is an invalid description of how they will be maximally exploited going forward in this century. Instead, as pointed out by (none other than) Hubbert in 1975, these flux-limited resources will tend to follow a logistic production growth curve such as that depicted in Figure 5.6 (Hubbert in Ruedisili and Firebaugh 1975).

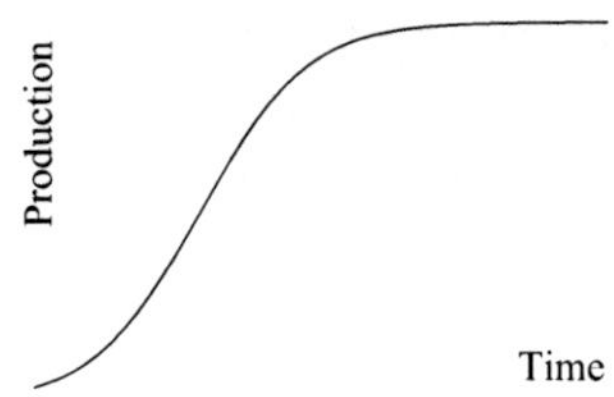

Figure 5.6. A logistic growth curve showing exponential growth and saturation.

In logistic growth, production will initially grow exponentially and once it approaches some physical, economic, social, or environmental threshold, growth will slow and stabilize to some constant production value. This should make sense, because many of the flux-limited resources we have been developing for the past several decades, in particular hydropower and more recently wind energy are harnessed where there are large amounts of untapped, high-flux energy (areas of high wind speeds, or large natural water discharges). This will ensure rapid, exponentially increasing installed capacity and energy production since the best sites for these energy installations are still abundant and available. (This, indeed, is the case with wind and solar power currently.) But, as the best sites are developed and the next best get developed, and so forth, the rate of growth will slow until it is not cost-effective to continue installing capacity. Practical limits will force growth to terminate, but because the source of the energy (the sun or Earth's core) is not depletable by human activity, decline in production will not necessarily occur, as long as the infrastructures are adequately maintained. Since so little energy is currently produced from the non-hydro "renewable" energy sources compared to annual fossil and nuclear energy production, it is plausible to expect exponential growth in production to occur for many years to come.

There are some additional details specific to certain popular resources that further limit their development, which are described briefly below.

5.4.1 SOLAR ENERGY LIMITS

Solar energy is probably the resource with the least need to consider practical limits. With thousands of times more energy arriving on Earth's surface daily than we consume, it is numerically feasible for solar power to be a relatively unlimited solution. However, there are still practical limitations to consider. One is the land use argument. Since most large-scale solar development concepts involve constructing huge solar power plants, covering thousands of acres with solar collectors does still cause a disruption to local ecosystems. (And yes, even deserts are ecosystems.) Large-scale deployment of solar power facilities may affect local climates, by altering the heat uptake and reradiation in the area. Solar thermal projects can mitigate this challenge to some extent by having multiple uses for the land. To the extent that society moves toward more decentralized energy systems, though, this factor can be mitigated. There are many millions of acres of rooftops in the world. If these surfaces are employed to mount solar panels, generating electricity for the building's occupants and tying into a local mini-grid, they need not require significant additional land.

Solar energy has other limits to face as well. For many years, photovoltaics have been made from very pure silicon in a variety of forms. Whereas silicon itself is incredibly abundant, the processing plants will require billions of dollars of investment—and a good deal of energy to process and refine the silicon. Thus, while photovoltaics have a very good overall "energy profit," they still incur an energy deficit when they are being produced. This will be one of their limiting factors in being able to respond to an energy crisis. They will make it worse before they are able to make it better! There are alternatives to silicon-based photovoltaic cells, including cadmium telluride, copper indium selenide, both with and without gallium as an additional component.[5] Some of these are showing particular promise, but it is easy to imagine that some shortages of the components may emerge as solar photovoltaic technologies grow to the point of making millions of modules per day. This rate of manufacturing would need to continue indefinitely as the rated life of a PV array is about 30 years—and the global energy market is growing! This manufacturing will also require a great deal of energy input.

5.4.2 WIND ENERGY LIMITS

Wind can be considered much like solar. Its overall resource base is huge and will continue as long as Earth does. Land may become a limiting factor in the long run, as well as the raw materials required in their manufacture. Naturally, the production of millions of wind turbines will incur an energy (and raw materials) debt as well, in the near term. This factor is not particularly significant unless the wind turbine manufacture is being rapidly scaled up in response to an existing energy shortage. It is also plausible that unforeseen environmental concerns emerge as more wind generators are placed in proximity to populations and natural habitats, which also would stunt its growth. It seems reasonable to think that placing several thousands of wind turbines across the world could impact local wind and temperature patterns, which could potentially cause disruptions. But those concerns should be considered as very long-term limits. For the near term, limits are emerging in the form of resistance to building wind turbines in many prime locations because of aesthetic arguments.

5.4.3 HYDRO ENERGY LIMITS

Large-scale hydropower, the kind that was aggressively exploited throughout the last century, requires locations that have large water flows, with either significant localized head (e.g. large waterfalls) or the potential to create a large head by impounding water behind a dam on the river or bay, to create a large instantaneous head at the dam. This has already been seen to impact ecosystems: totally disrupting local ecosystems in the flooded dam basin and altering ecosystems along the river and even at the mouth, as silt transport is radically altered. While the environmental impacts themselves will be discussed in the following chapter, the concern over environmental and social disruption limits the potential for new "big dam" projects. There is some real question about the extent of additional hydropower development that makes sense in North America and Europe, as well as about appropriate scales for any new hydropower development.

[5] A more complete description of the technologies can be found in Godfrey Boyle's *Renewable Energy.* (Boyle 2004).

5.4.4 GEOTHERMAL ENERGY LIMITS

The ultimate limits of geothermal power production are not clear at this point, although there certainly are limits, and it will take time after prices have remained high for some period of time to test and prove new technologies to expand the use of this large and relatively renewable resource base. Currently, the best locations for large-scale geothermal applications are located near the Ring of Fire (described in Chapter 3), where hydrothermal resources are being exploited. Electricity production from geothermal energy is limited to relatively shallow, high enthalpy reservoirs, with some form of natural permeability. Whether the rock is porous, like a sandstone, or has an extensive network of natural fractures, it contains water at very high temperatures, that can flow through the pores or fractures, much like oil and gas do and will produce to the surface as high-quality steam.

Indeed, the best sites are almost all being exploited in some fashion. But geothermal energy has a vast resource base, mostly untapped. Earth's interior not only contains a tremendous amount of heat, but is also generating more heat through natural radioactive decay, thus it has a potential to be used sustainably for a very long time. It is a resource that has both substantial stocks upon which to draw and substantial flux, or renewal rate. To date, only reservoirs at depths up to about 10,000 feet (3,048 meters) are tapped. Technologic advances, especially in hot dry rock geothermal power production have the potential to expand the portion of the resource base which can be tapped. There is much more hot rock at accessible depths than there are natural steam reservoirs. We would suggest that sustained higher prices for energy will facilitate the development and application of hot dry rock technologies. The surprising success of drilling long laterals, accompanied by massive hydraulic fracturing in the oil and gas shale plays seems to support the notion that technology can rise to the challenge of moving fluids effectively through vast rock formations with little to no intrinsic permeability. What could limit the exploitation of the geothermal resource is if drilling begins to cause unacceptable seismic disruptions.

5.5 WHAT DOES ALL OF THIS MEAN TO SUSTAINABILITY?

The goal of sustainability is one of ensuring adequate resources both for the present and for the future. In this way, sustainability cannot be divorced from its roots of sustainable development. This means that energy use must be extended to the more than three billion people who currently do not have sufficient energy to support survival, let alone development. Consequently, a piece of the transition-to-more-sustainable-energy puzzle must be continued growth of energy supplies for several decades. The physical limits of petroleum resources make it unlikely that oil and gas will be able to continue to expand their *production rates* sufficiently to meet that challenge. However, the nature of a petroleum peak and decline is likely to be much more gradual than is popularly portrayed. Prices are likely to be high, which will incentivize more aggressive exploration and recovery technologies. A great deal of oil and gas remain in the world to be discovered and to be produced. The limit will be the rates, not the ultimate recovery.

Other depletable resources may be able to play a constructive part in the transition. It will be important to realize, however, that expanded use of any finite resource significantly shortens the time horizon until its production peaks. As for biomass, a long-term sustainable rate of use is possible, so long as it is less than the growth rate and does not adversely impact other critical systems.

Conservation is likely to be critical to achieve a sustainable global energy system. Conservation, though, cannot significantly curtail the benefits derived from energy use. For the Developing World, conservation can only mean efficient expansion of energy sources in the near term. Since every energy source has some form of limitations, it will be essential to move into the transition as efficiently as possible. Similarly, it will be important to use our resources to their greatest effect, which requires an understanding of energy resources as part of a global system. Oil and gas, then, are "alternative energy sources" to fuelwood dependent people. As depletable sources, they cannot be counted on indefinitely, but neither can they be realistically discounted in planning energy transitions to support sustainability—and sustainable development.

Without sufficient energy inputs, development is not possible. It may be argued that some of the affluent world may be better to adopt a steady-state economics than to continue on a quest for unlimited growth (see Daly 1991). There can be little argument, though, that the less affluent half of humanity needs economic growth, which can only be fueled by more energy. Recall that the Bruntland Commission definition, so frequently referenced by sustainability advocates (including your authors) is one of "sustainable development." The world will continue to need energy to meet its development needs. The pathway must preserve energy supplies for the affluent, while expanding the availability of energy to the world's poor.

Oil shortages, as for any commodity, are a function of two things: supply and demand. The thing that makes petroleum different from other commodities is that energy serves an essential role in human development and survival, with petroleum products (oil and gas collectively) providing the lion's share of the world's energy supplies. Food is similar to energy in that it is essential (and the similarity holds up in that much of its consumption in the affluent world is discretionary), but there is no single food source that so dominates food supply as oil and gas dominate energy supply. In so saying, note that, while crude oil and natural gas are different products, with different costs and benefits, they are results of the same natural processes, are produced by the same industry, and commonly occur together in the same reservoirs. Thus, oil and gas can be viewed as parts of the same resource base, which currently provides more than half of the world's energy. How long can these limited resources continue to meet the growing demands of a growing population?

REFERENCES

Akins, J. 1973. The oil crisis: This time the wolf Is here. *Foreign Affairs* 51: 462–490. doi: http://dx.doi.org/10.2307/20037995

Ayres, R. U. 2007. On the practical limits to substitution. *Ecological Economics 61* (1): 115–128. doi: http://dx.doi.org/10.1016/j.ecolecon.2006.02.011

Bartlett, A. A. (1978). Forgotten fundamentals of the energy crisis. *American Journal of Physics*, *46* (9): 876–888. doi: http://dx.doi.org/10.1119/1.11509

Boyle, G. 2004. *Renewable Energy.* New York: Oxford University Press.

Campbell, C. J. 2002. Forecasting global oil supply 2000–2050. Hubbert Center Newsletter 2002/3: 1–7. hubbert.mines.edu/news/Campbell_02-3.pdf

Daly, H. 1991. *Steady-State Economics*. Washington, DC: Island Press.

Deffeyes, K. S. 2010. *When Oil Peaked.* New York: Hill and Wang.

Deming, D. 2000. "Oil: Are we running out?" Second Wallace E. Pratt Memorial Conference, San Diego.

Deming, D. 2003. Are we running out of oil? *Policy Backgrounder* (159): 1–14.

Ebenhack, B. W., and S. Wilson. 2005. "Clarifying the Limits of Petroleum." Annual Meeting Proceedings. New York: AIChE.

EIA (U.S. Energy Information Administration). 2011. *Annual Energy Review*. http://www.eia.gov/totalenergy/data/annual/index.cfm

Grossling, B. F. 1977. A critical survey of world petroleum opportunities, *Project Interdependence, Library of Congress*. Committee Print 95-33: 645–658.

Hern, W. M. 1999. How many times has the human population doubled? Comparisons with cancer. *Population and Environment 21* (1): 59–80. doi: http://dx.doi.org/10.1007/BF02436121

Hirsch, R. L. 2005. Shaping the peak of world oil production. *World Oil 226* (10): 61–65.

Hubbert, M. K. 1949. Energy from fossil fuels. *Science* 109: 103–109. doi: http://dx.doi.org/10.1126/science.109.2823.103

Khodeli, I., ed. 2009. *From Curse to Blessing? Using Natural Resources to Fuel Sustainable Development*. Hoboken: Wiley-Blackwell.

Kunstler, J. H. 2006. *The Long Emergency.* New York: Grove Press.

Masters, G. M., and W. Ela. 2008. *Introduction to Environmental Engineering and Science*. Upper Saddle River: Prentice Hall.

Maugeri, L. 2004. Oil: Never cry wolf-Why the petroleum age is far from over. *Science 304* (5674): 1114–1115. doi: http://dx.doi.org/10.1126/science.1096427

Maugeri, L. 2006. *The Age of Oil: The Mythology, History, and Future of the World's Most Controversial Resource*. Westport: Praeger.

Meadows, D. H., and Meadows, D. 1972. *The Limits to Growth.* New York: Signet.

Ruedisili, L. C., and M. W. Firebaugh, eds. 1975. *Perspectives on Energy*. New York: Oxford University Press.

Ruth, M. 2006. A quest for the economics of sustainability and the sustainability of economics. *Ecological Economics. 56* (3): 332–342. doi: http://dx.doi.org/10.1016/j.ecolecon.2005.09.012

Schollnberger, W. E. 2006. From scarcity to plenty. Who shapes the future mix of primary energy? What might it be? *Oil and Gas European Magazine* 32: 8–20.

Simon, J. 1996. *The Ultimate Resource II.* Princeton: Princeton University Press.

WWI (World Watch Institute). 2010. Oil Consumption Hits All-Time High. http://www.worldwatch.org/oil-consumption-hits-all-time-high

CHAPTER 6

Environmental Impact

This chapter focuses on the environmental component of sustainability as it relates to energy choices.

6.1 THE ENVIRONMENT AND HUMANS: INTERCONNECTED SYSTEMS

The word "environment" refers to the totality of that which surrounds us: nature, resources, the world itself, and other humans, are all part of our environment. Humans impact the environment and the environment impacts humans as well. The prevalent focus on the concept of environmental sustainability stems from the reality that humans depend on natural inputs to survive and to *sustain* a good quality of life, but overuse of those inputs can lead to unwanted and *unsustainable* degradation for humanity and the world.

Throughout most of history, we humans have viewed ourselves as contending against the environment. Nature was something to "tame," to "defend oneself from," and "to have dominion over." The natural world was threatening in many ways: it was filled with more powerful predators; its temperature swings could be fatal; and it occasionally brought us terrifying natural disasters. The recognition, though, that we are a part of the environment and are dependent on nature for our food, for pure water, and for clean air, has gained broad acceptance in recent decades. There is a general awareness that we must find balances between our exploitation of natural resources and preserving them for the future.

Human activities invariably have some impact on the natural environment: both in terms of the resources extracted from the environment and the waste streams returned to it. When human populations were small, our impact was minimal. However, human endeavor has been a story of more people using more resources to do more things, thus creating more impactful wastes along the way. Those wastes contaminate our land, water, and air both at local and global levels.

Environmental impacts are commonly characterized as "externalities." They are external to the economy in that they are not directly reflected in resource pricing structures. Of course, externalities also include impacts on people, and the impacts can be either positive or negative, but it is most often the negative impacts that are discussed. Indeed, a major basis for environmental concern is often the impact on people. Some negative externalities that have very direct human impacts include the impacts on human health from air and water pollution, which are directly linked to human consumption and waste production. Moreover, the scale of our consumption demands larger scales of operations, which incur the risk of ever-larger impacts.

6.1.1 THE ENERGY AND ENVIRONMENT FOCUS

Energy draws particular environmental attention for several reasons. Perhaps the most conspicuous is the large waste stream of combustion products released to the atmosphere by burning carbon-based fuels. Particulate matter and oxides of carbon, nitrogen, and sulfur are the most significant environmentally damaging products of combustion on a local scale, while carbon dioxide (CO_2) has become more closely scrutinized in recent years for its potential for severe and detrimental global impacts. Likewise, spills in producing and transporting petroleum can have significant environmental consequences. A great deal of petroleum pollution ultimately contaminates the oceans from the unheralded but daily minor spills from fueling vehicles, changing oil, and even the escaping vapor from pumping gasoline that eventually condenses out with rain water (NAS 2003). Nuclear power, of course, has radioactive waste products; not only the much-discussed high-level waste from spent fuel, but also the low-level radiation in the massive reactor vessels and containment facilities. Finally, little popular attention is given to the waste streams from the admired "renewable" energy sources, but most solar photovoltaic cells, for example, contain gallium and arsenic: toxins that ultimately find their ways into the waste stream in the manufacturing process as well as at the end of a cell's life.

In the sections that follow, we will discuss general topics on impacts and then specific impacts from the energy resources we exploit every day to maintain our modern way of life.

6.2 CHARACTERIZING ENVIRONMENTAL IMPACTS

Environmental impacts can come in a wide variety of forms, thus the need to characterize them, at least in general terms. There are numerous references to understand each type of impact better (Masters and Ela 2008; Ristinen and Kraushaar 2005; and Graedel and Allenby 2010 are particularly useful), but for the scope of this text, we deem it necessary to ensure that readers can have a basic background in the issues.

6.2.1 TOXINS, POISONS, AND TOXICITY

Toxins and poisons are those substances that directly harm living cells or tissues, and are widely transmitted through the air that we breathe and the water that we drink. Toxins most specifically refer to organically produced poisons, and toxicity can be used to assess the quantity and severity of exposure to living things. Some substances are highly toxic to most living cells in small doses (e.g., strong acids and bases, chlorine, or hydrogen cyanide). Other substances are only slightly toxic, like alcohols, sodium chloride (salt), carbon monoxide and even small particles, to name a few. Some toxins are acute and others are cumulative. Acute toxins can kill quickly, but their effect often diminishes quickly when the source is removed. Cumulative toxins, as their name suggests, build up in tissues with increasing damage. In excessive concentrations, everything can be hazardous. Anything that displaces some of the oxygen from air can result in asphyxiation, but this is not toxicity. Even too much oxygen can be hazardous.

6.2.2 RADIATION

Radiation is discussed frequently as an environmental and human health hazard. It is most often associated with nuclear power, where radiation is endemic to the fuels and to the process. However, the fly ash from coal-fired power plants also contains significant amounts of low-level radiation. Radioactivity appears to be one of the most misunderstood of the environmental hazards, eliciting a response as if it is much worse than the chemical hazards from other energy system waste streams. The fear, no doubt, stems from the potential for unparalleled levels of contamination and hazards to humans from the rare, but real catastrophic releases in the case of major accidents at nuclear power plants, or from nuclear weapons.

Attention has focused more tightly on real potential nuclear hazards in light of the disaster at Chernobyl in 1986, and the more recent Fukushima disaster in 2011, after an earthquake and tsunami cut off cooling water to and damaged four nuclear reactors. The appeal of nuclear energy is that small quantities of purified fuel produce vast quantities of energy and there is enough prospective nuclear fuel to extend our access to electricity for many decades. It also has no carbon emissions, so ameliorates the trend toward amplifying global climate change. The concept of risk, not only to workers, but also to the general public, has significantly constrained its development in the United States, while fears of nuclear weapons technology proliferation have (and likely will continue to) inhibit its deployment to meet the needs of the Developing World.

6.2.3 HUMAN SAFETY AND WELFARE

Since humans are a part of the natural environment, as well as architects of the built environment, it is appropriate to address safety and welfare impacts, especially since the acquisition of energy entails hazards to both.

The Macondo blowout in the Gulf of Mexico draws attention to the hazards of drilling. There is no question. Drilling involves very large equipment, generating tremendous force to cut through miles of rock—rock that contains fluids at very high pressure. Careful application of safety protocols greatly diminishes the risks. But people make mistakes. They fail to notice warning signs. They try to save a little money (indeed, often it is a lot of money, but it is always "a little" relative to the disaster it can prevent). Equipment fails, even when engineers design "redundancy" into systems. When pipe ram number one fails to seal the well, you engage pipe ram number two—or the Hydril, or the blind rams. Redundancy reduces risk, but cannot eliminate it. There may be a failure in part of the system that does not have redundant safety mechanisms. Critical problems can also occur outside of the crisis containment technology—a bad cement job around casing in a drilling well can allow gas to move up the annulus, outside of the casing and thus outside of the blowout prevention system (OSC 2011).

But oil and gas operations are by no means unique in presenting hazards. Underground coal mining remains a particularly hazardous occupation . As stated earlier, nuclear power plants can fail, through human error or through natural disasters impacting facilities. And when the scale of operations is very large, the scale of the calamity associated with a failure is comparably large. Small scale, though, does not eliminate risk. Many small-scale activities are likely to be totally unregulated and unable to employ sophisticated equipment. This means that frequent, small-scale activities combine to present large overall risk. A woman in Africa dying of respiratory disease from cooking over open fires four hours a day will not reach the newspapers: In part because it happens every 16 seconds! The scale of the global crisis is monstrous, in spite of the individual events being small.

Also, people seldom discuss traffic fatalities in the context of energy and sustainability, but we argue that it is worth noting that the form in which we consume one eighth of the World's oil in the United States leads to some 40,000 American deaths annually. It is not a cost of the production or delivery of the energy, but it is a cost of how we consume. The World Health Organization estimates that global traffic accidents constitute one of the three largest killers of people between ages 5 and 44, with about 1.3 million annual deaths as of 2010 (WHO 2011). That is a staggering death toll. We would certainly not accept it in most other contexts. The consumption choices are a part of the overall system—and a part over which the average person has some control. To the extent that we choose to live closer to work, school, and activities, we choose to reduce the number of traffic fatalities. To the extent that we choose to deploy and employ public transportation, we choose to reduce traffic fatalities. To the extent that we choose not to use vehicles over-sized for their purpose, we choose to reduce traffic fatalities.

Finally, it should be clear that everything we do has risks. The more we drill, or mine, or run power plants, or go to the woods to gather fuel, the more accidents there will be and the more deaths will be attributable to each energy system. Mitigating risk, to some extent, means reducing overly consumptive activities. Every time that you decide to stay home and read (or play a game, or even talk to your family), there is effectively zero probability that you will be killed in an auto accident. Every time you choose to get in the car, that risk becomes nonzero. In a similar vein, the choice to consume automatically brings with it the need to produce, transport, or process more energy—all of which incur risks for people and the environment—even for "green" energy systems.

6.2.4 LAND USE AND ECOSYSTEM DISRUPTION

Rich ecosystems have provided humanity with great riches throughout history. From natural resources like timber and rubber, to the many pharmaceuticals derived from plants and insects in tropical jungles, ecosystems have been a vital piece to our prosperity. The intricacy of the balance of ecosystems makes disruption particularly challenging to mitigate. Certain species depend on very specific other species for their sustenance—or even reproduction, as some plants are pollinated by specific species of insects. When one piece of the ecosystem is removed, it affects others and an imbalance can result.

Large-scale operations of all kinds require land and other resources. When land is transformed to human use, it carries with it some disruption of ecosystems, whether through flooding, or clearing for crops, or contamination. For instance, roadways disrupt ecosystems, as evidenced by the road kill that can be seen in nearly any day's driving. Likewise, one of the most common critiques of solar energy relates to the land required for very large-scale energy production.

Yet, humanity alters land for its uses in increasingly substantial ways. These alterations have resulted in a cultural transition from living in rural settings to increasingly urban settings.

This has in turn resulted in a number of strains on resources like water and air, as well as land. Congestion, pollution, and sometimes complete eradication of species have been some of the consequences of a more prosperous world. Since modern science has yet to study in depth 99% of the species found on the planet, there is a strong probability that we are depriving ourselves of many wonderful—and valuable—discoveries as we impinge on the rich diversity of species (WCED 1987).

6.2.5 WATER USAGE AND POLLUTION

Water resources are used in a variety of energy systems: thermal power plants use water to create steam; it is needed in irrigation for growing biofuels; and, rivers are impounded or diverted for hydropower systems. None of these uses take the water out of the surface hydrologic cycle, or make it unpotable. Nevertheless, the changes inflicted on the water system can severely affect local ecosystems.

The vital water resource system is also impacted by discharges and spills: Oil spills being the most notorious of these. They can impact the local ecosystems dramatically at the time of the spill. The extent of long-term ecosystem effects needs to be adequately characterized. There seems to be some impetus to conduct such studies in the aftermath of the 2010 Macondo blowout in the Gulf of Mexico.

Water is used in energy production processes and contaminated by waste streams and accidents. Acidification from the emissions of oxides of sulfur and nitrogen, as well as CO_2 can form acids in the atmosphere. When acids are deposited over bodies of water or are carried into the water through runoff, they can lower the pH, making the water unfit for some species of fish.

Many energy production processes consume water. It is not consumed in the sense of being lost—or transformed into something other than water. Water is a very stable chemical compound, it is very rare to transform it to something else, but the water can be polluted, depriving it of its great value—suitability for drinking and supporting aquatic life. The water can be contaminated in many ways, including simply becoming saline. It may also become less accessible to humans or to natural ecosystems through the way it is used or disposed of.

A great deal of water is injected into oil and gas reservoirs, either for the purposes of enhanced recovery or fracturing the formation around well-bores to permit better initial flow rates (frac jobs). Much of the water injected into subsurface petroleum reservoirs is lost to the surface hydrologic cycle for a very long time, as opposed to water use in agriculture. On the other hand, it is worth noting that oil and gas energy systems ultimately release a great deal of water to the surface hydrologic cycle. Consider that the oxidation of hydrocarbons yields energy plus CO_2 plus water. Indeed, there is a molecule of water produced for every molecule of CO_2 released to the atmosphere. Paleo plant life took up CO_2 from the atmosphere and used energy from the sun to combine it with water to form organic molecules, which ultimately became oil and gas.

Therefore, the production of fossil fuels actually increases the amount of fresh water on Earth's surface, by a much larger factor than the amount of water consumed by injection. Does this mean that the water usage is a nonissue for petroleum? No, there can still be local imbalances created and a good deal of the water vapor produced by burning fossil fuels will end up in the oceans. There are multiple factors to consider with the use of water in oil and gas operations, as is true with most of the issues. Context is always important. If the water is drawn from long-term storage in aquifers, its use impacts future availability of water from those aquifers. If the water is contaminated with toxins, local ecosystems can be seriously imperiled.

Water is often consumed in producing energy. The impact on the water resource base can be highly variable. In much large-scale agriculture (from which biofuel feedstocks are derived), large quantities of water are commonly drawn from subsurface aquifers. While the water stays in the surface hydrologic cycle, it is certainly possible to deplete the store of water in the subsurface aquifers. Power plants and large industries have, historically, discharged warm water into streams and rivers, which can impact local ecosystems.

6.2.6 AIR EMISSIONS AND POLLUTION

According to the United States Environmental Protection Agency (EPA), emission "is the term used to describe the gases and particles which are put into the air or emitted by various sources." Thus, air pollution is, in a sense, using the air—or more accurately, the atmosphere—as a dumping ground for emitted waste streams. The sources of these emissions primarily come from combustion fuels, and the amount of emissions entering the atmosphere varies annually, "caused by changes in a nation's economy, industrial activity, technology improvements, traffic, and other factors" (EPA 2012).

At its basis, combustion requires three essential elements to proceed: 1) a fuel; 2) an oxidizer; and 3) a heat source to initiate the reaction. Using a generic chemical expression, combustion proceeds in the following way:

$$\text{Fuel} + \text{Air} \rightarrow \text{Heat} + \text{Airborne Emissions} + \text{Ash}$$

Note that oxygen in air is the only chemically active compound contributing to the combustion reaction. Thus the use of air, which contains about 78% nitrogen, serves to decrease the efficiency of the combustion process. Combustion of fuels such as biomass, coal, oil and gas, inevitably emits some products of incomplete combustion, known as air pollutants. Common air pollutants include:

- Ozone
- Particulate matter
- Carbon monoxide
- Nitrous oxides (NO_x)
- Sulfur dioxide (SO_2)
- Lead

According to the EPA, of the above air pollutants, "particle pollution and ground-level ozone are the most widespread health threats."

Ozone is not a direct combustion product, but rather the product of sunlight induced chemical reactions between NO_x and volatile organic compounds (VOCs), which come from industrial and transportation pollution sources (electric power plants, automobile exhaust, gasoline, etc.). While ozone at high altitudes is desirable for UV protection, ground-level ozone is characterized as a pollutant and can be quite harmful to human health, because exposure can trigger a number of negative respiratory responses (EPA 2012).

Particulate matter (PM), on the other hand, is a direct combustion product, partly a function of incombustible contaminants in the fuel. Smoke is the product most commonly associated with PM, but indeed this class of pollutant is comprised of a mixture of particles and liquid droplets, typically very small in size (Reist 1993). Particle size and shape are directly linked to their potential for causing health problems, and environmental agencies worldwide are most concerned about particles that are 10 micrometers (microns or μm) in size or smaller. These fine and ultra-fine (2.5 μm or smaller) particles generally bypass the body's built-in defenses in the nose and throat, and once inhaled, these particles can affect the heart and lungs function. See Figure 6.1 for a comparative view of these small sizes.

Note that while we tend to associate airborne smoke and PM emissions with combustion from the fossil fuels, for some irrational reason, we now tend to neglect wood combustion emissions as a major pollution source. Raw biomass resources are particularly dirty, especially in terms of airborne pollutants. Indeed, the impacts on human health from firewood dependence are staggering. The particulates in smoke can cause or aggravate respiratory disorders, with smoke from indoor wood fires in developing countries being recognized by the World Health Organization as the seventh greatest killer in the world, and on par with malaria-related deaths on an annual basis (WHO 2012). As the world considers climate change (discussed below) as a leading driver for a transition away from fossil fuel consumption in the developed world, it should consider that changing from wood consumption even to fossil fuels in the Developing World would go a long way in improving environmental conditions and overall quality of life for more than two billion people. In biomass-dependent regions, a transition to fossil fuels (methane and propane in particular) can even slow deforestation, protecting an important carbon sink and ecosystems alike.

Finally, of the remaining common pollutants described in this section, we note that carbon monoxide is a local pollutant and poisonous even at small concentrations, NO_x is both a primary pollutant and an important reactant in the formation of ground-level ozone, while SO_2 is a leading cause of acid rain pollution.

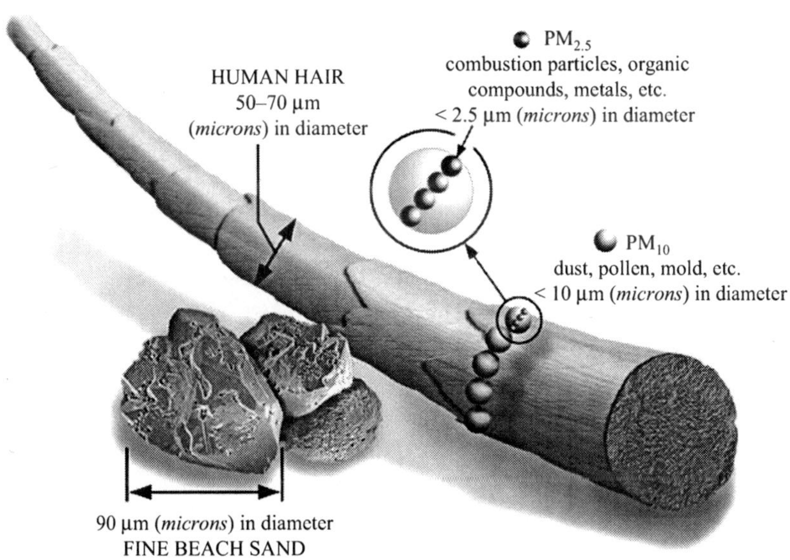

Figure 6.1. Particulate matter size and shape comparison.
Source: EPA (2012).

6.2.7 GREEN HOUSE GAS EMISSIONS AND CLIMATE CHANGE

One additional emission that results from the combustion process is CO_2. It is a compound essential to life on Earth, but its concentration has increased significantly since the beginning of the industrial era, primarily because of land use changes, industrial agricultural practices, and massive fossil fuel and biomass consumption. CO_2 also is a known greenhouse gas. It absorbs infrared radiation to help keep the planet cozy for life to exist. As concentrations of CO_2 increase, though, the planet may warm more in response to this excess amount. It is also possible that its presence will enhance the effects of other changes in the climate. As this is a global phenomenon, increasing concentrations of CO_2 in one location will increase the concentration everywhere. Thus, this is a global concern.

The Intergovernmental Panel on Climate Change (IPCC) was founded in 1988 as a joint venture of the World Meteorological Organization and the United Nations Environment Programme to investigate the potential impacts this phenomenon might have globally (IPCC 2008). Yet it was former United States Vice President Al Gore's 2006 documentary *An Inconvenient Truth* that propelled the issue into the greater popular consciousness. The issue became as highly polarized as it was popularized. It has been a rallying point for environmentalists to contend with pro-growth advocates. Numerous articles can be easily found contending that climate change is a flawed or even fraudulent theory. To many environmentalists, though, it has become the central issue for sustainability.

As we are not climate scientists, it is not our intention to take up the merits or detriments of the climate science debate in any detail. There are a number of other sources better suited to that (see Ruddiman 2010 and Tomkiewicz 2011). However, from a purely technical, objective perspective, we believe that there are a few points worth making about the issue.

First, climate change is a theory, by definition, just like relativity and evolution. Being a theory does not make a concept tenuous, but it does mean that it is subject to new information and on-going evaluation. Beyond that, climate change is based on numerical models. Models offer a means to characterize exceedingly complex phenomena. Earth's climate system certainly qualifies as exceedingly complex. The very nature of models is that they cannot be perfectly accurate. Models are employed to simplify systems that are too complex to evaluate with direct calculations on a deterministic number of measurements. The so-called skeptics are correct that the models are imperfect. So what? By the nature of models, we can predict with great confidence that the model's predictions are not perfect—just as a model that predicts your longevity will almost certainly not predict the precise moment of your death. Nevertheless, you will die: probably closer to the time predicted by a model than to a time derived in some more arbitrary manner. The only way we will be able to determine with absolute accuracy the actual climate change phenomenon is to wait until it has played out and generated sufficient data to prove (or disprove) key parts of the model.

Second, the technical basis of the models is beyond reproach. CO_2 is emitted by burning carbon-based fuels. The concentration of CO_2 is increasing in the atmosphere. CO_2 absorbs infrared radiation, which should produce a warming effect in the atmosphere. None of that is in question. The debates rage about the severity of outcomes predicted by the model. Do the models predicting dire consequences from anthropogenic carbon emissions sufficiently account for natural feedback mechanisms (Earth system responses to minor changes)? Do the models account for large, external effects, such as those that produce natural warming and cooling cycles on Earth? Climate change has drawn so much attention because it has the potential to be catastrophic to the environment and to people. If the climate change predictions are right—even sort of right, it is prohibitive to wait to see the results. No matter how skeptical some people may be of the evidence for climate change, it behooves us to take measures to mitigate its potential severity. By comparison, the likelihood of any given drilling well suffering a blowout like Macondo is extremely small, yet it is not only prudent, but an absolute requirement to take measures to mitigate against a worst-case scenario.

Third, most measures to ameliorate climate change will have salutary effects on energy conservation and environmental preservation. Indeed, more sustainable energy systems will almost certainly have lower "carbon footprints." Combustion fuels, which release CO_2, generally emit traditional pollutants and they invariably have stocks on which we draw—thus stocks which can be depleted—and are being depleted by current energy systems. The resource limits discussed in the last chapter correlate most strongly to environmental problems in the cases where the resource is a part of current ecosystems: biofuels. To some extent, climate change is a bell-weather for other environmental problems. The more CO_2 a combustion fuel emits, in general, the more polluting it is.

Finally, our primary critique of the climate change debate is that it has become *the* defining environmental issue. The focus on climate change, to the exclusion of other problems, allows unbalanced analyses that advocate for some energy alternatives simply on the basis of *claims* of lower greenhouse gas emissions. Other forms of pollution, we believe, receive too little attention now, as do land use, water use, and other sustainability issues. Some authors try to advocate for corn ethanol, for instance, by saying that CO_2 emissions from ethanol do not count, because of a purported "closed carbon cycle" for biomass. This argument claims that carbon emitted by burning biofuels is taken up by the next round of vegetation growth. If we accept that argument at its face, we are drawn away from the other problems of ethanol production from corn: land use, soil depletion, monocropping, effects on biodiversity, water use, and harmful by-products to name a few. When we critically examine ethanol production, though, we must not ignore the time it takes for vegetation to regrow and what else might be growing on land if it were not growing corn for ethanol. If the land in question were abandoned parking lots, the carbon cycle would be truly closed. More likely, though, the land would otherwise be forest, grassland, or other kinds of productive cropland, in which case, the annual growth of biomass was already part of that system. Thus, the growth of vegetation is not directly related to combustion of corn products as a fuel source (Ebenhack and Martínez 2008).

We suggest that it is imperative to consider the potentially catastrophic risk of climate change, but that we should not fixate on this issue, to the exclusion of others. Energy systems have complex sets of costs and benefits, all of which should be considered in concert.

6.3 ENVIRONMENTAL IMPACTS OF THE SOURCES

Every energy source has some impacts associated with various stages of the production, conversion, and end-use of the energy. It is tempting to view certain activities or impacts in isolation. For a time, electric cars were overly hyped as "zero emissions vehicles." At the point of use, this was quite accurate, but it ignored the emissions at the power plants generating the electricity to charge those vehicles. In reality, the environmental impact needs to include the impacts of the additional (coal or nuclear) power plants and their fuel use. It can still be argued that the impacts would be less than those from millions of exhaust pipes from cars running on internal combustion engines. Nevertheless, the real impacts, across the full energy system need to be considered. When we move on to the point of viewing global energy systems collectively, it is desirable to optimize the collective impacts from all of the energy systems together with the benefits provided by each. A variety of different smaller impacts are likely to be less of a burden on the global environment than a few much larger impacts.

6.3.1 COAL

As of the writing of this book, coal is generally acquired either through underground mining or through open pit "strip" mining. (We consider "mountain top removal" as a particular kind of strip mining for the purposes of this text.) While advances in the affluent world's regulatory requirements and technologies have greatly mitigated the risks of underground coal mining, it remains a very dangerous career. The incidence of fatal injury is lower than a number of other occupations, but incidence of deaths from what is commonly called "black lung disease" adds hundreds of fatalities per year—down from a couple of thousand per year within the last 40 years (NIOSH 2011). Underground mining continues to take a toll on human life expectancy. Immediate hazards in the underground mines include roof collapse, flooding (from breaking into an underground aquifer), explosions, and fires. Some underground coal fires have burned for decades, inevitably contributing atmospheric pollution (the mine must be connected to the air, or else it would not have oxygen to support combustion.)

Both underground and strip mines can contaminate aquifers. Water moving through the mine can leach out a variety of contaminants, including heavy metals, organic materials, and carcinogens. A student assistant a few years ago, who was from Dimock Pennsylvania (where much of the controversy around "fracking" Marcellus Shale gas wells is centered) points out that coal mining went on for decades in the region. He suspected that some of the contaminants being found in the area waters seem more likely to be related to old coal mines than to the frac jobs. Whether or not his speculation is accurate, it is true that coal mining operations have resulted in contamination of water resources. Both regulations and modern technology can, and do, work together to mitigate this risk, but can probably not eliminate it altogether.

Surface disruption is highly visible for strip mine operations. All of the overlying soil and rock are stripped away into tailings piles at the edge of the pit. This exposes the coal seem to be completely removed. The process alters the surface topography, which changes rainwater drainage patterns and can lead to both erosion and to contamination of surface waters. Modern operations typically call for restoring the surface topography to some close semblance of the original, ensuring that the topsoil is on top, and replanting vegetation. This can mitigate the long-term effects, but the near-term ecosystem is still disrupted and surface waters may risk contamination during operations.

Once the coal is mined, it is transported to the consumer. In modern times, the consumer is generally a power plant or large industrial facility. Power plants often employ dedicated unit trains that run strictly from the mine to the plant—typically 100 cars in each unit train, each car carrying 100 tons of coal. Barges are also often used to move coal, if a river is nearby. Since the coal is massive, there is an energy (and thus environmental) cost in the transport, which is a function of the distance. The enormous masses involved also incur some human hazards.

The conversion facility burns the coal either for process heat or to generate steam to produce electricity. Modern facilities are increasingly employing cogeneration, in which electricity is generated, along with process heat. This increases the benefits derived, for the same energy and environmental costs. Burning coal, of course, produces significant air pollution. Coal produces more carbon dioxide per unit energy than any other fossil fuel. It also produces more particulates and other forms of pollution. Large-scale facilities have the opportunity to employ the best mitigation technologies, though. Coal combustion also produces a great deal of solid waste. Fly ash piles accumulate some of the worst toxins, including heavy metals like mercury and radioactive materials. Perhaps, as regulations apply more value to externalities, there will be incentive to "mine" the fly ash to recover these materials and use them.

Clean coal technologies have the potential to mitigate all of the environmental costs (externalities) associated with coal. There are many technologies taking on the "clean coal" mantle, but they are not all likely to be effective. When energy prices reach and maintain higher levels, these technologies may be sufficiently deployed to compete effectively with each other—and with the other energy options.

6.3.2 OIL AND GAS

Natural gas is the cleanest burning of the naturally occurring combustion fuels, followed by fuels derived from oil. Natural gas is predominantly methane: one carbon molecule attached to four hydrogen molecules. Upon burning, it produces CO_2 and water—and very little else. If you doubt the cleanliness of burning gas, imagine cooking a meal in a kitchen without a chimney using any other combustion fuel. (It would probably kill you to try it, so just imagine it.)

Methane is a powerful greenhouse gas, so leaks from natural gas contribute disproportionately to greenhouse gas pollution. However, good engineering protocols can virtually eliminate leaks. Natural gas always has and always will leak from geologic reservoirs from whence it escapes to the atmosphere, through natural processes. The gas that we produce and consume, to some extent, reduces the natural leakage to the atmosphere. Nevertheless, since the quantities of gas being produced annually is immense, the amount of gas escaping from leaking storage vessels, flanges, pipelines and test procedures will dwarf the reduced natural emissions in the near term. We consider leakage to be a serious problem, requiring good engineering practices throughout the world, but do not believe it should significantly restrict gas development.

Crude oil is typically the preferred hydrocarbon commodity, especially on the global markets. It is much easier to load a great deal of dense crude oil into large tanker vessels than it is natural gas. Gas that was produced with the oil has often been flared, because there was no local market. This is still true in some locations remote to local markets. One of the most perverse examples of this is in Nigeria, where the people in onshore communities, such as Port Harcourt, could certainly use the energy, but the markets are not sufficiently developed, where more than a billion cubic feet of gas are flared every day (EIA 2011). This is also a function of companies and governments not working together effectively to find local markets for gas.

The gas is being flared because it does not have an adequate market, but the companies still want to produce and sell the oil internationally. Flaring is wasteful and environmentally damaging. In poor communities, it also fuels local resentment. More aggressive action needs to be undertaken to reduce and stop flaring. It would be very costly to capture all of the flare gas in such places as Nigeria and develop local markets for it. There must be a commitment at least to start with capturing some of it and delivering it to local consumers. This sort of waste will rightly raise significant environmental and social critique. The industry and the petroleum engineering profession must put a high priority on finding ways to eliminate such waste, especially in close juxtaposition to such deep need.

The most widespread concern about gas at this time in the United States, relates to the technologies used to produce some of the vast accumulations of shale gas. The Marcellus formation in the eastern United States has been found to be a rich reservoir, which can be produced economically through the combined technologies of horizontal drilling and extremely large hydraulic fracturing (frac) jobs. The fracs are raising extensive environmental concerns. Fracturing does consume a great deal of water: often more than a million gallons per well. The frac fluids do contain some chemicals to control the fluid's viscosity and to carry solid materials (generally sand) to prop the frac open, among other things. The vast majority of the frac fluid is water and sand: 99.5%. About 10% of the water is produced back shortly after the frac job. In response to environmental concerns, most companies have begun to recover essentially all of the water produced back, treat it and reuse it in the next frac job. Of course, some of the chemicals in small concentrations are produced back and treated, while a large part remains indefinitely in the formation, miles beneath the surface.

A large concern relates to contaminating potable aquifers with the frac fluid, or with gas. Pictures of a man lighting water from his kitchen faucet, shown in the documentary film *Gasland* have become popular images. Geomechanical analyses indicate that it is virtually impossible for an induced fracture to extend upwards through thousands of feet of overlying rock to break into shallow, potable aquifers. (We say "virtually impossible" because, as engineers, we are both aware of how many factors one may not account for, such as the possibility of an induced fracture merely connecting to an existing natural fracture, which in turn reaches the aquifer. Nevertheless, it is highly unlikely.)

Did the fellow in the movie try to light his kitchen faucet before drilling came to Dimock? (If so, why?) We would suspect that the results would have been the same. It is much easier to envision the physics of gas moving through natural fractures over a million years to establish a small saturation in aquifers than for an induced fracture to connect through thousands of feet of rock and mobilize gas in a matter of weeks or months to establish a saturation in this fellow's water well. (But, has it not been observed at several locations, not just the one in the movie? Yes. That makes the explanation of small amounts of gas migrating to the aquifer over geologic time even more reasonable.) That's how gas reservoirs are formed. The process of gas migration into aquifers over geologic time is well known. Tests have begun to evaluate the presence of gas in aquifers, by testing for gas dissolved in water in regions where no drilling has yet occurred. We have not yet seen published results.

There are documented cases of inappropriate completions procedures contaminating water. State agencies identified several cases and are compelling remedial work and compensation from the companies that did poor casing and cementing jobs. If the wellbore is not properly sealed with casing and cement, it can provide an effective conduit for gas or oil to reach the surface and near-surface aquifers. That is why regulations are in place around most of the world to set surface casing and cement across all known potable aquifers when drilling for oil and gas.

A study published by the National Academy of Science has shown strong evidence of methane-contaminated aquifers in the vicinities of a number of producing gas wells. This study quite properly assesses several mechanisms to explain the contamination. While many areas have found that surface waters and aquifers contain some methane from natural, biogenesis from decay of contemporary biota, the study ruled out this possibility for their set of wells. They did, however, conclude that the most likely mechanism was leaking casing and cement, rather than the hydraulically induced fractures (Osborn et al. 2011). Indeed, their map showing the increased probability of contamination with proximity to the surface location of the wellbore, offers very strong support for this conclusion. Shale gas wells commonly have very long horizontal sections, reaching a mile or more laterally from the surface location of the well. If the contamination resulted from the frac jobs, the contamination would not be expected to center on that surface location.

One of the things companies should do to ensure environmental responsibility is to test the cement bond on the surface casing frequently and thoroughly. Environmental activists can pressure regulatory agencies to ensure this kind of testing. The environmental groups and the responsible companies and professionals can work together to ensure that this most serious risk is properly addressed.

Oil is not as clean-burning as gas, but is still cleaner than most fuels. The way we consume gasoline, to support the proliferation of individual automobiles, especially in America incurs a very high cost in terms of human—and animal—fatalities. The transition to increased sustainability must, we believe, include a shift in consumption patterns, toward more public transportation and fewer numbers of cars on the roads.

The internal combustion engine and the automobile are not a very efficient combination. Barely 15% of the chemical energy of the fuel finds its way to the work of moving the vehicle. If you consider that the typical automobile is really serving the purpose of moving 50–200 kilograms of people, and a few kilograms of stuff, then the efficiency is even lower. The metric used is fuel use per "passenger mile" or "per ton mile" for transport of goods. Vehicular transportation is a major source of both air pollution and CO_2 emissions.

Oil spills are the other major, sometimes highly visible environmental impacts associated with petroleum. As mentioned previously, oil finds its way to contaminate the oceans from several sources. On average, spills from offshore drilling and production operations represent the smallest category, hardly accounting for more than 3% in most years. Transportation spills account for four times more oil reaching the ocean, with a large part of that coming from normal transport operations. Nearly half of petroleum in the sea comes from natural seeps. The next largest share is refined products from consumer spills, fueling vapors, etc. Still, when a major incident occurs, such as the Macondo blowout in 2010 or the Exxon Valdez tanker spill in 1989, the effects are dramatic and highly concentrated. There is severe short-term impact on the local ecosystems. In places where there is a great deal of marine birds and mammals (both of which spend significant time on the surface), the impacts are particularly stark. Whereas these disasters are rare, the environmental consequences are terrible. Professionals working in petroleum exploration, production and transport must be well trained, not only in the technologies to be employed, but in the responsibilities we have to society—and to the world. Human errors can be very costly, both to ecosystems and to human life.

6.3.3 NUCLEAR

Nuclear power, which was championed as a clean, safe, and peaceful energy alternative immediately after World War II, has been met with much resistance in the past 40 years, and indeed does not adequately meet all of the conditions put forth in defining a sustainable energy source. Nuclear power is an energy system with extraordinary potential, but its development is limited largely by public perceptions of its risks. It appears that some people conflate nuclear power with nuclear weapons. While they work on the same principles, they are really quite different. There are real environmental problems associated with nuclear power, but they are neither the threat of vaporizing square miles around the power plant, nor of the reactor core "melting through Earth to China."

The reactors generate some 5,000 gallons (19,000 liters) of highly radioactive waste each year for a typical gigawatt power plant and many times more low-level radioactive waste (Walker 1983). There has been no global consensus reached on how best to handle the radioactive wastes. Most nuclear power plants have undergone renovation and recertification, which also helps forestall the problem of what to do with the massive structures contaminated with low-level radiation.

Half-lives are often cited when speaking about the alarming potential of nuclear disasters. We would suggest that this could be misleading. Certainly, some of the fission products and original fuels have extremely long half-lives. But, many chemical toxins are permanently dangerous. The fact that we can calculate how quickly radioactivity decays does not make it more dangerous.

Still, the disasters at Chernobyl in 1986 and Fukushima in 2011 warrant significant attention. We must always seek to learn from mistakes, to avoid being doomed to repeat them. Chernobyl had several design flaws (including the lack of secondary containment) and the crisis emerged through a string of human errors. The postmortem analysis is not complete at this time for the Fukushima disaster, but one lesson that may be learned is not to concentrate hazards in one location. Six large reactors were constructed in a location that seemed to have little impact on any local populace—minimizing the not-in-my-backyard (NIMBY) factor. However when the earthquake and tsunami struck that location, the disaster was multiplied by the scale of the overall facility. Ever larger scale is not desirable.

Concerns over the proliferation of material and technology for nuclear weapons are likely to prevent nuclear power from being a solution for much of the Developing World. Even with proposals for self-contained reactors to be buried on location and run until they are spent, the black-box nature of such a design is an obstacle to human intervention if a problem does emerge. We think that engineering design can address the safety and environmental concerns about nuclear power in affluent nations (when the need for energy overcomes fears), but that it is not likely to be deployed in much of the rest of the world.

6.3.4 THE "RENEWABLES"

Renewable energy has been touted as the most likely to meet the sustainable energy development criteria. Renewable fuels can be superior to the fossil fuels in respect to resource depletion criteria. Our use of solar and wind does not deplete their availability for the future. However, the use of biomass at rates that exceed renewal rates, or carrying capacity of the local soil and water resources can draw down their stocks—and damage the local ecosystem balance at the same time.

Some of the renewables will be important parts of the transition to sustainability, but the conflation of renewability with being environmentally friendly is problematic. Literally, there is no necessary relationship. The term "renewable" references the rate at which something is renewed or replenished. Intrinsically, this has nothing to do with the environmental impacts of an energy system. Environmental impacts are functions of the acquisition, the use and the waste products generated.

Whereas biomass is renewable in a human timeframe and can be anthropogenically renewable through replanting, the impacts of fuelwood dependence are quite severe. Wind has drawn a great deal of negative environmental attention for aesthetic reasons as well as concerns about hazards to birds and bats. The recognition of environmental problems with large-scale (big dam) hydropower seems to have led some authors to strip hydropower of its renewable label—which is absurd, because the hydrologic cycle is based on solar power and gravity, neither of which is being depleted by human use. (The other possible reason to deny hydropower its renewable status is that dams do silt up and need maintenance, eventually leading to a limited lifespan for the dam. Of course, though, the equipment deployed for any energy system requires maintenance and ultimate replacement, so this argument also seems specious.) Assessing environmental impact will demand more than simplistic categorization: it's not just a matter of renewability.

6.3.4.1 Hydropower

Large-scale hydropower projects alter the flow of rivers. Large basins are flooded, displacing local people (often farmers) and radically changing local ecosystems. The river slows nearly to a halt, forming a large lake or reservoir upstream of the dam. Different species live in lakes than rivers. The river no longer transports silt and sand to deltaic estuaries at the river's mouth. Even the prevention of periodic flooding means that the soils surrounding the river are not being replenished by its sediments. The dam also impairs fish migration. Whereas "fish ladders" were built at a number of large hydropower dams, there is some evidence that fish migration is being disrupted. Perhaps the fish spend too much of their bodily chemical energy, searching for the new path around the dam and cannot reach their spawning grounds.

The scale of the big dams also invokes the risk of catastrophic failure. If a large dam were to be breached, the huge impoundment of water in the reservoir would suddenly overflow the downstream river channel, potentially flooding out many communities.

In general, environmental interests tend now to oppose new big dam projects, as do local residents in many cases. We suggest that smaller scale, run-of-the-river projects may have greater potential, as they create much less disruption. These smaller hydro projects may also be well suited to the smaller energy demands of communities in Developing Nations, and fit well with distributed or decentralized energy.

6.3.4.2 Wind and Solar

Wind and solar probably have the smallest "environmental footprints" of the energy alternatives. Nevertheless, it is important to acknowledge those they do have. Photovoltaic solar systems commonly employ toxic "dopants" to facilitate the flow of electrons. These toxins may present hazards if solar photovoltaic power scales up to the point that many solar cells are failing and going to the dump at the same time. Land use is often mentioned in regard to solar and wind power, but it is wise to compare the land use to other energy systems. Coal-fired electricity, for instance, uses land for the power plant, the fly ash storage, and in strip mines (if that is how the coal is obtained.) Furthermore, both wind and solar power have potential to be developed in combined use formulations. Rooftop solar has considerable potential. Wind farms need not preclude other activities.

Bird (or raptor) kill was once a serious concern for large-scale wind turbines, but these were turbines whose blades spun at high revolutions per minute (RPM). Modern wind units tend to be low RPM, high torque. We have not found credible evidence of *significant* bird kills associated with the modern, low speed turbines. Watching these modern turbines at work, it is rather difficult to imagine birds being struck by the slow-moving blades.

Solar energy is tapped in ways beyond the popular photovoltaic cells. Solar thermal electric generation avoids the toxins used in making solar cells, but does use water and chemicals to prevent corrosion, freezing, and algae growth in the working fluids. Some of these chemicals will doubtless leak to the environment, with some negative consequences. This is not expected to be a large factor.

Whereas we include human health and safety in the environmental impact section, we do not include aesthetic concerns. They cannot be totally ignored, as they are part of the social acceptance of an energy system, but we suggest that they are too subjective to analyze in any robust way in regard to environmental impact.

6.3.5 BIOFUELS AND BIOMASS

Half of the world depends on raw biomass for its meager energy supplies: generally in the form of cooking over open three stone fires. The World Health Organization estimates that the pollution from these wood fires kills some two million people annually, from a variety of respiratory diseases (WHO 2012). This represents a catastrophic human health impact, many times larger than from all of the nuclear accidents to date—and it happens every year. It also affects many young children, who are with their mothers for hours each day in smoke-filled cookhouses.

Wood fires are probably the most polluting of any energy source. They only seem romantic in a context in which they play a very minor and occasional role—and in which the biomass is abundant relative to demand. We believe that the transition to more sustainable global energy systems must entail the opportunity for the poorer half of humanity to move away from wood, charcoal, and dung. In these contexts, oil and gas are clean, "alternative" energy sources. Indeed, remote wood collection and transport, charcoal markets, and dung patty production depict the harsh reality of dependence on raw biomass.

Modern biofuels are also combustion fuels, with all of the emissions of the fossil fuels and raw biomass. Some claim that the "closed carbon cycle" for biomass takes up all of the CO_2 produced when burning the fuel. As discussed earlier in this chapter, we suggest that it is fatuous to ignore all of the other environmental impacts of any energy source and use carbon emissions as a single surrogate for environmental costs.

Nevertheless, let's look into the closed carbon cycle for biomass argument a bit further. It is based on several assumptions. First, is a time scale assumption: that regrowth occurs on nominally the same time scale as the consumption. At some level, of course it does not. We raise questions about the practical validity of this assumption. The enormous disparity between the time frame of accumulation and conversion is the basis for concern about carbon emissions from the fossil fuels: millions of years of accumulation being released to the atmosphere in a matter of years. By the same reasoning, though less dramatically, the consumption of wood from mature forests can releases centuries of carbon storage every year.

More importantly, the closed carbon cycle concept implicitly assumes that the new growth results from the use of the biofuel. If something else could be growing there (whether producing food, or fixing carbon for relatively long periods in trees, or even grasslands), then we suggest that the closed carbon cycle is based on a false premise. The true carbon cycle analysis must take into account all of the energy and material fluxes into and out of the system. If a good deal of gasoline or diesel or coal is used in growing and processing the biofuels, they must be considered in the carbon balance. Finally, CO_2 is not the only environmental impact. We believe that the land use, the water requirements, the external energy inputs, and the full range of carbon and pollutant emissions must be evaluated for each biofuel system.

6.4 COMPARING IMPACTS

How do we compare the widely disparate environmental impacts of different energy systems? First, we must be able to understand the range of impacts of complete energy systems. There are resource impacts, as discussed in the previous chapter. Depletion of fossil fuels is a resource impact, but not truly an environmental one. Whereas the depletion of forests from firewood demand or clearing land for agriculture is a major environmental impact, some land use incurs less environmental cost than others. If the land can have multiple uses, the impact is lower.

Land and water use are related impacts. These are both systems that play essential roles in the well-being of the environment. Different water uses can have varying levels of impact. Water can simply be used on fields and evaporate to rejoin the hydrologic cycle. If it has not been contaminated, the impact of such use is only significant to the extent that it deprives water from other local uses. (For instance, rivers that are entirely diverted to human use never reach the sea, destroying the downstream ecosystems.) Water used for irrigation of crops is likely to pick up some contaminants from the soil. If the fertilizers are organic, then the problem is largely eutrophication—fertilizers in the rivers and lakes promote algae blooms that can choke out other life forms.

If synthetic fertilizers, pesticides, and herbicides are used, the contaminants may be directly toxic. The suffix "cide" means that the chemical is toxic, at least to certain kinds of living things. Water used for cooling thermal power plants can alter ecosystems, when it is returned to rivers at higher temperatures. Water used in massive hydraulic fracturing jobs (becoming common in the development of gas and oil production in shales) is lost to the surface hydrologic cycle indefinitely. Of course, much more water is released to the environment as a result of the immense scale of global fossil fuel combustion processes.

The main products of combustion, as discussed previously, are the infamous CO_2—and water. Our current global energy system is actually adding water to the terrestrial environment. Combustion of petroleum releases a molecule of water for every molecule of CO_2. Whereas the frac jobs on shale wells may use a few million gallons of water, the combustion of fuel from a well producing one million cubic feet of natural gas per day will deliver one million gallons of "paleo"[1] water to the atmosphere every 50 days. The use of water from confined aquifers (those which are not significantly recharged from the surface) still constitutes depletion of valuable

[1] By "paleo," we mean that many millions of years ago, growing plants used solar energy to combine carbon dioxide and water to form organic molecules. The fossil fuels represent some portion of that ancient photosynthetic biomass that has been stored through geologic time.

underground water storage. The overall water balance on the surface of Earth, though, increases as a result of the water of combustion.

One of the challenges in planning energy transitions is the fact that every energy system brings its own, unique sets of environmental costs. There are some commonalities within certain classes of energy systems. All of the combustion sources produce CO_2 and airborne pollutants. Every energy source has an impact on land use. But the various energy systems incur a wide range of negative externalities. How do we compare them to make decisions? Is it better or worse to increase radioactive wastes to reduce greenhouse gas emissions? How do we compare the effects of oil spills to altering river ecosystems and sedimentation with big hydropower projects? How should aesthetic issues be weighed?

Ecosystems can cope with assaults until a critical threshold is passed. To the extent to which we put all our eggs in one or two energy baskets, we are also concentrating our environmental costs. This factor also suggests that there are costs to extremely large-scale projects, for which we have not adequately accounted. These projects concentrate the impact. Scale of operations should be considered as carefully as the category to which the source belongs.

Planning an energy portfolio may require a balance that minimizes overall damage to the environment. This will require both understanding the impact vectors (such as pollution, radiation, greenhouse gases, human health factors, etc.) and understanding their individual and collective impacts.

REFERENCES

Ebenhack, B. W., and D. M. Martínez. 2008. "Old Carbon vs. New Carbon." Society of Petroleum Engineers Annual Health, Safety and Environment Conference, Nice, France.

EIA (U.S. Energy Information Administration). 2011. *Country Analysis Briefs: Nigeria*. http://www.eia.gov/emeu/cabs/Nigeria/Full.html

EPA (U.S. Environmental Protection Agency). 2012. *Particle Pollution*. http://www.epa.gov/airquality/particlepollution/

EPA (U.S. Environmental Protection Agency). 2012. *Urban Air*. http://www.epa.gov/air/urbanair/

Graedel, T. E., and B. R. Allenby 2009. *Industrial Ecology and Sustainable Engineering*. Upper Saddle River: Prentice-Hall.

IPCC (Intergovernmental Panel on Climate Change). 2012. http://ipcc.ch/organization/organization.shtml

Masters, G. M., and W. Ela. 2008. *Introduction to Environmental Engineering and Science*. Upper Saddle River: Prentice-Hall.

NAS (National Academy of Sciences). 2003. *Oil in the Sea: Inputs, Fates, and Effects III*. http://www.nap.edu/openbook.php?isbn=0309084385

NIOSH (National Institute of Occupational Safety and Health). 2012. *Occupational Respiratory Disease Surveillance Database*. http://webappa.cdc.gov/ords/norms.html

Osborn, S. G., A. Vengosh, N. R. Warner, and R. B. Jackson. 2011. Methane contamination of drinking water accompanying gas-well drilling and hydraulic fracturing. *Proceedings of the National Academy of Sciences*. *108* (20): 8172–8176. doi: http://dx.doi.org/10.1073/pnas.1100682108

OSC (Oil Spill Commission). 2011. *Deepwater: The Gulf Oil Disaster and the Future of Offshore Drilling*. http://www.oilspillcommission.gov/final-report

Reist, P. C. 1993. *Aerosol Science and Technology*. New York: McGraw-Hill.

Ristinen, R. A., and J. J. Kraushaar. 2005. *Energy and the Environment*. Hoboken: Wiley.

Ruddiman, W. F. 2010. *Plows, Plagues, and Petroleum: How Humans Took Control of Climate*. Princeton: Princeton University Press

Tomkiewicz, M. 2011. *Climate Change: The Fork at the End of Now*. New York: Momentum Press.

Walker, C. 1983. *Too Hot To Handle*. New Haven: Yale University Press.

WCED (World Commission on Environment and Development). 1987. *Our Common Future*. New York: Oxford University Press.

WHO (World Health Organization). 2011. Factsheet No. 292. *Indoor Air Pollution and Health*. http://www.who.int/mediacentre/factsheets/fs292/en/

WHO (World Health Organization). 2011. *Global Plan for the Decade of Action for Road Safety 2011–2020*. http://www.who.int/roadsafety/decade_of_action/plan/en/index.html

CHAPTER 7

Global Social Contexts

In this chapter, we discuss energy's role in development and its social implications in the 21st century.

7.1 MODERN ENERGY'S ESSENTIAL ROLE

In Chapter 6, we focused on the environmental implications of energy use. However, in addition to the many implications energy use has on the environment, there are equally important social implications that will become even more explicit and complex as humanity progresses through the 21st century. This is so because many of the resources we harness to maintain our high functioning, technology-driven society are transported from what are known as "resource-rich, cash-poor" regions. Despite being responsible for massive energy flows into global economies, the citizens of these undeveloped regions are often forced to deal with poor social infrastructures, which can be traced almost exclusively to lack of appropriate energy infrastructures. So as humanity has been focusing on reconciling modern energy demands with the need to preserve the biosphere (Smil 2003), an emerging and perhaps even more pressing issue is addressing how to manage production and consumption patterns globally to cope with expected and needed increases in demand in poor regions, while simultaneously improving consumption practices in wealthy nations to meet acceptable, equitable, and more sustainable consumption rates.

Indeed, approximately half of the seven billion people on Earth at the time of this writing still rely on raw biomass for primary energy. Where there is biomass dependence, there is pervasive poverty. The same half of the world's population constrained to firewood, crop residues, and dung live on less than \$2 a day (WRI 2004). If a country's per capita income is less than \$300 per year, approximately 90% of the population depends on raw biomass, whereas once income exceeds \$1000 per capita, most people use modern fuels (Rajagopal and Zilberman 2007). We suggest that growth in income depends at least in part on access to a greater quantity and/or better quality energy.

In developing countries, even those who have connections to the electric power grid often find it erratic and unreliable. In Nigeria, Africa's most populous country and the continent's second largest petroleum producer, electricity supply in urban areas averages just 8–10 hours per day.

At any moment, some three quarters of the nation's power plants are not operating. Grids may be so poorly maintained that less than 60% of power generated is sold to consumers, with the rest being stolen or lost to inefficiencies. Stealing electricity may seem odd, but desperate people do take desperate measures. A trip into the countryside of a developing country is likely to show wires (sometimes bare) running from power line to houses, while businesses and affluent residences have inefficient and polluting diesel-burning generators to augment unreliable grid service. In his article entitled "Toiling in the Dark," Michael Wines notes the relationship between grid stress and World Bank estimates that the lack of reliable power has depressed annual economic growth rates by more than 2% in some of the world's poorest economies (Wines 2007).

As discussed in Chapter 6, dependence on biomass may represent a more egregious challenge to the environment than fossil fuels. Lest one forgets, fossil fuels are merely old biomass, which has actually been concentrated and "refined" by nature. Regardless of the age of carbon-based fuels, they have similar sets of combustion products. Raw biomass has the added challenge that it is drawn from current ecosystems and its use may place significant stress on the natural environment. The forests, which comprise approximately 24% of global land area, are being lost in many developing countries, at least in part due to firewood demands. At a conference in Ghana in 1993, the recently retired Dean of the University of Ghana-Legon, rose from the audience after a presentation by Ebenhack and said, "When I was a boy, you could not see for the forests, now you can see as far as you want." Dr. Orocco-Tete went on to explain that he attributed much of this to the rising demands for firewood. Deforestation is still driven significantly by growing numbers of poor, who depend on biomass for their cooking needs. Thus, biomass dependence drives a vicious cycle of poverty. Its use exacerbates the very poverty that makes people dependent on it. As it drives deforestation, it is robbing people of their environmental quality, of ecosystem stability, and of goods and services the forests can provide.

Deforestation rings approaching 12 miles (20 km) have been noted around some African urban centers. This is about the distance that people can walk to gather fuelwood. Beyond this distance, firewood markets, such as the one depicted in Figure 7.1, as well as charcoal markets develop, bringing vast quantities of forest to the cities to burn. Around rural villages, it is not

Figure 7.1. One of several firewood markets in Ouagadougou, the capital of Burkina Faso.
Source: Photo Courtesy of J. Mathewson and AHEAD Energy Corporation.

uncommon to see small caravans of women, carrying burdens of firewood as large as themselves. On a variant of Macbeth's premonition, the forest doth fall and come to Ouagadougou and to Accra and Maputo and Addis Ababa.

The Food and Agriculture Organization (FAO) estimates that deforestation accounts for one quarter of all man-made emissions of the greenhouse gas (GHG) CO_2. Destruction of forests adds almost two billion tons of carbon to the atmosphere each year. To the extent that the forests are degraded, there is no offsetting uptake of these carbon emissions. Preventing this stored carbon from escaping is important for maintaining the global carbon balance and vital to conserving the environment. Providing cleaner, more efficient energy options would relieve the pressure on forests and fix as much as 15 tons of carbon per hectare per year in forests rather than releasing it to the atmosphere (FAO 2005). In addition, cooking fires produce onerous indoor air pollution. Most cooking is done over simple 3-stone fires, which are extremely dirty. Women often cook for hours in poorly ventilated cookhouses. In addition to the impacts of smoke on women, the soot constitutes a global climate change menace as well. Some estimates suggest that soot from open wood fires in the Developing World may contribute 18% of the overall climate change mechanism (Ramanathan 2008).

The health consequences of biomass dependence are staggering. It leads to illness and death for many people: especially women and children. The pollutants from cooking over open fires (an activity which commonly takes several hours every day) lead to millions of deaths every year. Some put the fatalities at 1.6 million per year, while others suggest that it is more like 4 million per year (Lim et al 2012). Most of the victims are women and their children, who spend hours with their mothers in the smoke-filled cookhouses. This virulent form of indoor air pollution kills more people than malaria (1.2 million per year), and almost as many as unsafe water and sanitation. It is the fourth greatest cause of death and disease in the world's poorest countries. The smoke contributes to asthma, lung disease, cancer, cataracts, tuberculosis, as well as still-born and low-birth-weight babies (Smith and Cowlin 2005).

Biomass dependence is especially hard on females because gathering fuel is seen as women's work in many cultures. Women and girls regularly carry as much as 20 kilograms of wood a few kilometers a day: This excessive stress loading results in debilitating musculoskeletal disorders. Furthermore, the task of gathering firewood is hazardous for those in rural settings, where it remains possible to collect wood. Cutting wood with simple machetes often leads to serious injury. The women and children gathering wood are at risk of attack by predators and venomous snakes. Sadly, the women are also vulnerable to rape and murder as they sojourn miles from home into the forests to gather the wood to cook the family meals.

In urban and periurban settings, the poor spend much of their income on energy: more than one-third of household expenditures in some countries. For the rural poor, energy-related activities consume a large portion of their time. The trek to the forests may take hours—or a day out of the work week. The time spent collecting fuel is time that cannot be spent in educational opportunities or other productive activities. Of course, these activities would require additional energy, from modern sources.

Despite this wealth of evidence for the need to help improve the circumstances of much of humanity, it remains difficult to convey modern energy's essential role in carrying out sustainable development strategies around the world. For the world's affluent, it is easy to recognize that reliable food and water supply, housing, transportation, communication, education, and medical care are essential components of a civilized world. It has been less easy to recognize, or perhaps, accept that these essential services are maintained by consuming sufficient amounts of modern energy in the form of fossil fuels and electricity. It should be obvious that as we plan to develop and implement new energy systems that aim to reduce environmental impacts, we must also plan to address the social needs that modern energy provides to regions that desperately need to improve their well-being. You simply cannot use traditional biomass to run modern industries.

7.2 ENERGY REQUIREMENTS TO MEET HUMAN NEEDS AND WANTS

Energy really is the key to economic and social development. And, once an energy infrastructure has been established for a given social structure, people within that structure "will naturally trend to higher levels of needs as they move from bare subsistence to being productive members of society" (Ebenhack and Ebenhack 2008). That is, if they have the opportunity and if they can afford it, people will naturally adopt a new energy system or energy systems that improve their quality of life. This progression up an energy/income "ladder" is a common theme when discussing household energy use, and is a microcosm for energy use as a whole. Figure 7.2 is an example of a typical ladder, and in it we see that at a basic level, people will use energy first for cooking, heating, and lighting, then add to that as they try to improve on those most basic

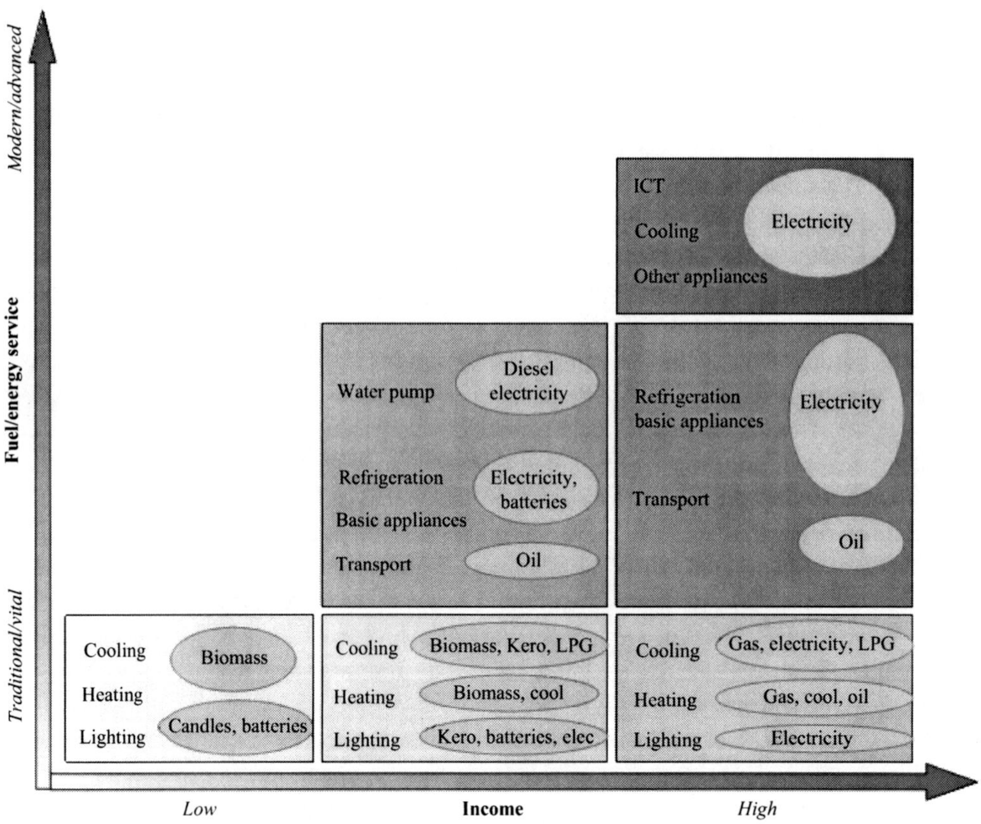

Figure 7.2. Illustration of a household fuel transition, otherwise known as an energy ladder. Note: ICT is information and communication technology.
Source: IEA (2002). Copyright [2002] IEA-OECD.

needs. At low incomes, biomass will be the dominant fuel type, and as income rises, modern fuels and electricity will be adopted.

Notice in Figure 7.2 that more modern fuels tend to be more versatile, more powerful and cleaner than traditional fuels. While there is some romantic appeal to wood fires in the affluent world, people do not balk at taking up demonstrably better energy systems. When a system is more effective, more convenient, and affordable, it may be adopted even more rapidly than supplies can support. For example, through our work in Africa, we were made aware of a project in Ghana in the early 1990s that sought to capture natural gas being flared at a local refinery. The gas was bottled and sold to local households, along with inexpensive stoves, for cooking. The Ghana National Petroleum Company went to some lengths to publicize and market the advantages of bottled gas over wood for cooking. Demand blossomed so rapidly that it outstripped the supply of gas and high pressure gas canisters. It was obvious that gas was a superior fuel.

7.2.1 HUMAN NEEDS

What is interesting in Figure 7.2 is that at its essence, fuels are assimilated on the basis of meeting vital human needs. But what constitutes a vital human need? In an effort to characterize needs and to identify how societies evolve in tandem with more robust energy systems, we borrow a concept from Abraham Maslow's work more than a half century ago, in which he set forth a hierarchy or pyramid of needs, which build upon each other. There are five different levels in Maslow's hierarchy of needs:

1. Biological and Physiological Needs—The most basic needs include: food, water, air, warmth, sleep and procreation. People are instinctively driven to these needs and cannot survive without them. Everything else is secondary until these needs are met.
2. Security Needs—Security involves protection from immediate threats, including: the elements, predators, enemies, and disease. These tend to be risk factors, rather than certainties, such as death from malnutrition.
3. Belongingness and Love Needs—Having biological and security needs addressed enable people to direct attention to love, friendship, and community.
4. Esteem Needs—Next, attention turns to having responsibilities, building personal worth and status, achieving social recognition and reputation.
5. Self-actualizing Needs—Finally, at the pinnacle of development, with all other needs sufficiently met, people can turn to personal growth and more spiritual needs.

Though there is ongoing debate about Maslow's concepts and how all these needs are interrelated and if they truly progress in this hierarchy, we find the idea of an evolving progression of needs helpful as we now turn to the question of energy's role in meeting vital human needs. For convenience, let us look at a common depiction of Maslow's hierarchy of needs side by side with a depiction of a hierarchy of fuels in Figure 7.3.

At the first and second levels (biological, physiological, and safety needs), food must be secured, distributed and processed; water must be obtained and often purified; dwellings must be found or built; and there must be a provision for lighting and useful tools. In addition to raw materials requirements, common fuel requirements to meet these needs include crop residues, dung, and wood, as well as charcoal, and plant and animal fats. Large numbers of people still live at this level, and only a small fraction get to meet their higher-level needs.

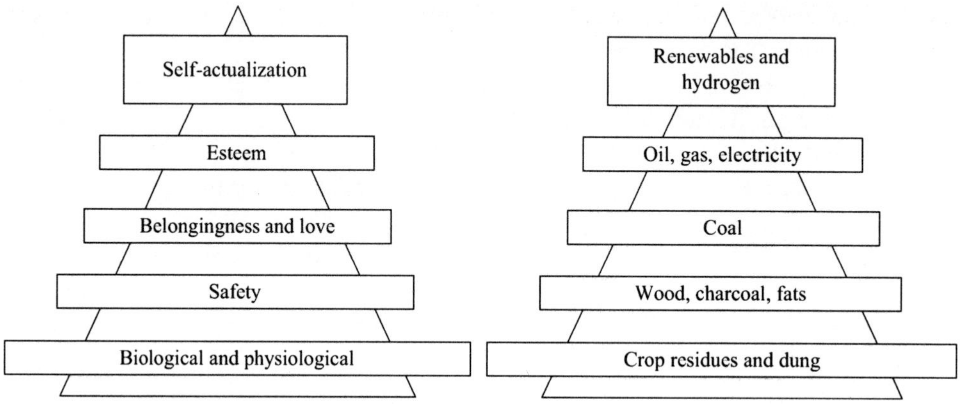

Figure 7.3. A common depiction of Maslow's hierarchy of needs compared to a hierarchy of fuels.

At the third level (belongingness and love needs), fulfilling social needs becomes very important. Activities to fulfill these needs include communication, transportation, and the building of public infrastructures. At this level, all of the "lower" fuels are used by all facets of society, while petroleum and electricity facilitate the next rung. With modern fuels, facilitation of these needs has progressed at a pace unlike anything seen in human history—and with cleaner emissions. These fuels represent high quality, abundant, affordable, and reliable energy in the Developed World, whereas in the Developing World, the availability of these resources benefit the very wealthy, but magnifies the discrepancies in achieving self-actualizing needs for the poorest members.

At the fourth and fifth levels (esteem and self-actualizing needs), quality of life becomes particularly important. There is a greater demand for public services in the form of schools and hospitals, and dealing with environmental degradation becomes important. At these upper levels, all energy options are on the proverbial table, but there is a concerted effort to adopt cleaner technologies, including from the renewable (or nondepletable) sources. It also opens the door to begin thinking about energy sufficiency. That is, thinking about whether you have enough energy to fulfill needs without continually increasing energy consumption.

The progression from bare subsistence to modern industrial and postindustrial societies requires ever more energy inputs as well as new technologies not only for household use but also for agricultural, industrial, commercial, transportation, and public services. Thus, increased demand for energy is a function of there being more and more people in the world as well as a function of taking up more energy to meet higher level needs.

7.3 THE ADVANTAGE OF CONSUMING ENERGY

Energy quite literally fuels both survival and development. Without energy, there can be no industry or technology. Modern agriculture owes its productivity to vast energy flows, while even the most primitive agriculture employed animal power (from humans before draft animals) to work the soil and harvest the produce. Life is a process of harnessing energy for our metabolic functions. Indeed, existence itself relies on energy. At absolute zero temperature (a complete lack of heat), matter ceases to exist.

It should be no surprise that energy is linked to affluence and poverty. Since it is essential to life, it cannot simply be done without. Chemical energy from food is required to support life: ranging from 1000 kilocalories per day for a relatively sedentary child to over 2500 per day for a physically active adult (roughly 250–1000 BTUs of chemical energy). Without this, bodily systems fail to function adequately. Generally speaking, food also needs energy inputs to produce, process, preserve, and/or cook. The minimum external energy input for basic cooking needs is estimated to be 50 kilograms of oil equivalent (kgoe, about one-third of a barrel) per capita per year (over 5000 BTUs per day) (UNDP 2005).

The move to industrialization went hand in hand with a massive increase in energy demands. Much of this progress involved replacing human labor and animal labor with machinery. Mechanization fostered tremendous increases in productivity, while the reduced demand for human labor made it possible for people to engage in creative, analytical, and management tasks. Mechanization expanded free time opportunities. The work week became shorter. This, in turn, enabled people in the working class to attend school. Literacy levels soared. Broader access to education, in turn, spawned even more innovation and new technologies. The process of technologic growth does indeed seem limitless. We must recall, though, that the resources on which development is based are not.

The advent of machines created demand for technical innovation. If machines could pump water from coal mines, run presses, or huge textile looms, perhaps they could do other things as well. Perhaps they could power vehicles: steamboats and locomotives were developed. The steam engine was the primary workhorse of the Industrial Revolution. It was found that, with some clever arrangements of valves and pistons, the pressure of steam generated by burning wood or coal could turn gears and gave rise to many great inventions. Each one taking more energy. Some inventions improved on the early steam engines, making them more efficient. Energy use took some steps backward and some steps forward, but the overall effect was continual increase in energy demand.

Improving technology in the Industrial Revolution did not provide uniform quality-of-life benefits for all of the people. The Revolution was in many ways a dark time in human development. The ability to replace labor made it easier to view workers with disdain. Many business owners found ways to use new technologies and the work that external energy sources provided in order to cut their payroll and build their wealth. Some of the tasks required in working with machines were found to be most readily performed by the small, nimble hands of children. Thus many youngsters were employed for pennies a day, working in factories and coal mines at considerable hazard. The new wealth created jobs that drew people in hoards. Some of this draw was because so many of the farmworkers were non-landowning peasants, with little chance to get ahead. The new wealth undoubtedly also created hope for many people that they could join the ranks of the *nouveau riche*.

Only in the 20th century did the Industrialized World see significant advances made in worker rights and protection against exploiting child labor. The balance between the freedom of capital and the rights and dignity of workers has been a struggle throughout much of the Industrial and perhaps post-Industrial ages. Harnessing massive quantities of energy has allowed many of us to move from subsistence livelihoods up Maslow's pyramid to achieve, or at least to aspire to, rewarding work—yet unequal access to these benefits remains one of the great challenges for sustainable development.

As energy-fueled mechanization has reduced the need for human labor, it has also provided the potential for extreme precision in work. This has eliminated the need for many tradespeople, craftsmen, and even artisans. This seems to have resulted in a loss of pride in craftsmanship: much of what we build today is not meant to last, but to be replaced with something even more cheaply made in short order. This has been one step on the path away from a sustainable society. The premise of a throwaway society is that there are no limits to the raw materials that go into making disposable stuff. We believe that a more sustainable world will be one in which products are built to last and the work of all is valued.

Energy also enables services as well as the production and delivery of goods. Consider a modern hospital: gleaming and brightly lit throughout. The tests that doctors can now perform to search for a myriad of ills are all energy consumers. An MRI can consume more than two kilowatts of power. If it is in frequent use, this is comparable to a common American household. Every instrument uses energy. All of the cleaning, scrubbing, and autoclaving uses energy. Recalling the embedded energy of materials, disposable items also represent significant energy consumption. Modern schools also depend on significant amounts of energy. It is practically assumed now that a school must have a vast array of computers, printers, and projectors in order to give students a viable education. (Even if we challenge that presumption, lighting does remain important to schools, and perhaps heating and cooling as well.)

Mobility may be the greatest transformation wrought by energy and the Industrial Age. It fundamentally changed how people live in affluent nations—especially the United States. The railroad was instrumental in the expansion of the United States into western territories and the conquest and subjugation of native populations. Consider that "Custer's Last Stand" came seven years after the dramatic linking of the Central Pacific and the Union Pacific, spanning the breadth of the continent with rail (and, as a frame of reference, 17 years after Drake's oil well). Pioneers and settlers certainly continued to use horses, oxen, and wagons, but their progress was supported by rail transportation and the shipping of goods—in both directions. The famous cattle drives, which were the subject of so many western movies, were to move cattle from the open ranges to the railheads, for shipment to meet the growing demand for meat in the east. Euro-American society spread out across the broad continent.

Interstate and intercontinental shipping also shaped modern affluent economies dramatically. Of course, even in the early colonial days, effective shipping was essential. Goods had to be moved great distances to provide the acquired wealth to the investors who supported colonization. Fleets of sailing ships operated for all of the colonial powers, culminating in 19th century clipper ships, which could travel more than 250 miles per day, speeding goods to customers. The advent of coal and then diesel ships did not significantly increase the maximum speeds of ships, but made the speed reliable. This boosted international trade to modern levels, where food goods are likely to travel more than 1000 miles (and perhaps several thousands of miles) to the consumer. This is at the heart of global trade, a "neo-colonialism" in which affluent nations encourage Developing Nations to produce certain raw materials, for export to affluent consumers.

It is evident that all of these fairly recent social developments are tied to economic productivity and underpinned by energy consumption. Energy is required to manufacture goods, to transport goods and people, to light schools and hospitals, and to provide a plethora of other services. These developments influence the energy system. However, the energy system, once established, also influences social and economic systems, as the energy sector itself promotes economic growth (Blok 2007). Economic growth, in turn, strives to improve quality of life. Following this trajectory, we see a vicious cycle of energy consumption leading to quality of life improvements, leading to even more energy consumption. There has to be a limit.

Understanding this energy/economy, or more appropriately, this energy/quality-of-life nexus is important for characterizing energy consumption in the 21st century.

7.3.1 IN-DEPTH: THE ENERGY/QUALITY-OF-LIFE NEXUS

Numerous researchers have sought the relationship between rates of energy consumption and quality of life, and a number of studies have found a positive relationship between economic conditions and per capita or total energy consumption at national and global levels. In particular, the correlation between annual total primary energy supply and gross domestic product is often used to indicate correlation. However, as Smil points out in his book, *Energy at the Crossroads*, this correlation "changes with developmental stages, and although it displays some predictable regularities there are also many national specificities that preclude any normative conclusions about desirable rates of energy consumption" (Smil 2003). Because of this reality, in more recent times a number of studies have tried to account for more of the factors attributing to quality of life. Of the many that have been considered, one seems to stand out: the United Nations Human Development Index (HDI). Despite the inevitable limitations to any metric that seeks to quantify a qualitative attribute, HDI is heavily employed in studying the relationship of energy use to quality of life. The United Nation's HDI is a function of an income index, an education index, and a life expectancy index—all normalized, so that the overall HDI metric ranges from zero to one (one being a perfect score). All of these factors are desirable, so HDI seems to be reasonably robust.

For a 2008 publication in the journal Energy Policy, we prepared a graphical comparison of the 2004 UN HDI to 2006 per capita energy consumption (PCEC, units of kgoe), depicted in Figures 7.4 and 7.5 (Martínez and Ebenhack 2008). Notice that the HDI increases rapidly at the lowest levels, with small increases in energy consumption, eventually leveling to a plateau. At

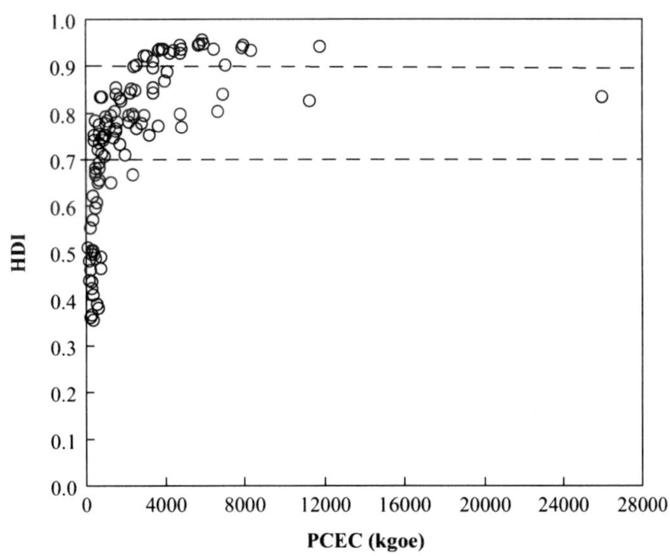

Figure 7.4. 2004 Human development index (HDI) plotted against 2006 per capita energy consumption (PCEC) for 120 nations.

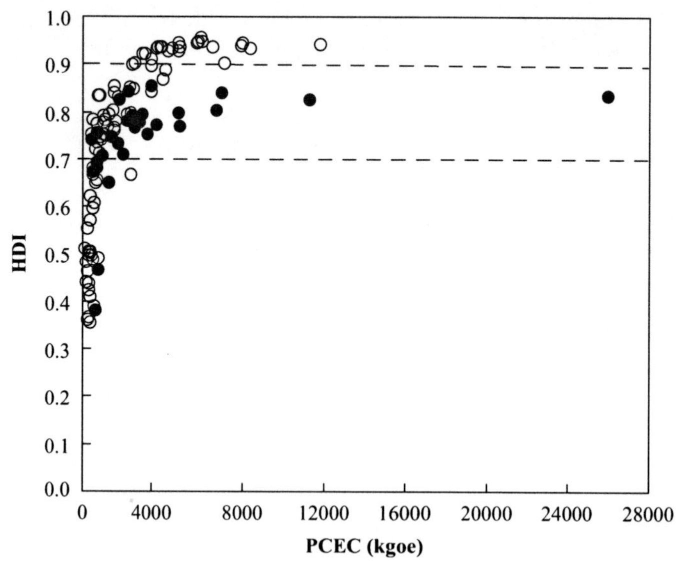

Figure 7.5. The 30 highlighted heavy energy–exporting nations in the HDI–PCEC curve.
Source: Martínez and Ebenhack (2008). Reprinted with permission from Elsevier.

the plateau, more energy ceases to add improved HDI. The general character of the relationship has been discussed by several authors (see Goldemberg 2001 and Smil 2003). Alan Pasternak (2000) used a logarithmic relationship to fit the data, suggesting a causal relationship between energy and quality of life. In that 2008 study for which the graphic was prepared, we found overall PCEC for each nation to show a noteworthy correlation with HDI, but there was some definite scatter in the data.

One point stood out to us, showing a high standard of living, but extraordinarily large energy consumption. It was not the United States, but Qatar. Granted, on a per capita energy basis this may not be a complete surprise, but we noticed that other OPEC and Former Soviet Union nations also diverged from the dominant trend. We reasoned that major energy exporters may have different factors affecting their energy consumption than the rest of the world. Applying a filter, separating nations with energy export-dependent economies from the rest revealed two separate trends, depicted in Figure 7.5. The two trends are parallel, with the major oil exporters reaching a plateau at a lower level of HDI.

Without the 30 energy export-dependent economies included, the primary trend is nearly definitive: for any given energy consumption, an HDI can be predicted with reasonable accuracy. In particular, the low-level consuming nations are shown in the near-vertical region of the open-circle curve (HDI = 0.35–0.7), representing much of Africa, the Indian Subcontinent, and Central America. The mid-level consuming nations are shown in the region of the curve that bends over (HDI = 0.7–0.9), representing much of Latin America and Asia. The high-level consuming nations are represented by the horizontal region of the curve (0.9 and above). These highest data points represent the Industrialized Nations. We dubbed the relationship between energy and human development the Energy Advantage. It seemed obvious.

In addition to the above observations, we were able to discern a few important points:

1. The low HDI nations do not consume enough to accommodate the most important aspects of development: mechanical power; refrigeration; lighting; transport; and communication.

2. Theoretically, given the steep nature of the Energy Advantage curve, little additional energy is required in low HDI nations to increase quality of life substantially.
3. An energy-export penalty seems to exist. The data show lower quality of life for export-dependent nations than those purchasing the energy.
4. Local capacity is important. The problem isn't that "energy poor" nations lack resources; it is the lack of capital and experienced professionals to develop modern energy resources for their own benefit.

More analytical work needs to be done to answer the question of causality, but there is reason to expect causality, at least up to an HDI of 0.8. It makes sense based on the vital role that energy plays in supporting most development activities. But there is mutuality. A growing economy needs more energy to support its growth, which growth also causes increased energy demand. The preponderance of evidence supports at least mutual causality. Energy is required for development and development increases the capacity to use more energy, in more forms. Energy does not make sufficient causality for development—it must be coupled with the services it supports: education, health care, business development.

The correlation between HDI and energy consumption mimics the saturation behavior of many natural processes. An easy analogy to relate to would be food. If one lacks sufficient nutrition, more food dramatically improves quality of life (and can improve life expectancy). There is, however, a saturation point, beyond which more food does not improve one's quality of life. Indeed, many of us in 21st century America have left the saturation point well behind and experience the reality that less food would actually improve our lives. Similarly, we believe that energy use is essential, to a point, beyond which its benefits diminish. This can be seen as good news. Some of the world's largest consumers, most notably in North America should be able to conserve energy significantly, without sacrificing any quality of life.

Finally, it seems obvious from the analysis that Africa is the region of greatest need: as reflected in terms of poverty and in terms of energy use. With about a billion people, Africa has a sixth of the world's population, but uses less than one-twentieth of the world's electricity. But the problem is worse than that. Of the small amount of electricity consumed on the continent, three quarters is used by South Africa, Egypt and a handful of other North African nations. Less than a quarter of people in sub-Saharan Africa have any plausible access to electricity. In the rural areas, where a large proportion of Africans live, 97% of the people may lack access to electricity. Excluding South Africa, whose economy and power consumption both dwarf their neighbors, sub-Saharan Africa's remaining 700 million citizens use about as much electricity as the 38 million citizens of Poland (Wines 2005). Much of the energy available goes to industry: a single aluminum smelter near Moçambique's capital of Maputo, for example, may use more power than the rest of the country combined. A vast majority of the people are left relying on excruciatingly little energy.

7.4 CONSUMERISM

For affluent nations, acquiring colonies allowed them to expand the resource base on which they drew and expanded their limits to growth. In modern times, this aspect of international trade remains. Less affluent nations trade their resources—including human resources—to provide the goods that permit consumer nations to continue economic expansion without suffering the environmental costs or resource depletion associated with their consumption. This gives a false impression that pollution is being overcome. Technology does enable us to do more with less pollution, but a significant part of the mitigation of environmental damage in affluent nations is a result of "exporting pollution" to the Developing World. Raw materials are extracted and sent to the West and, as nations climb a few rungs up the development ladder, they also begin to make steel and other partially processed goods. Indeed this phenomenon also masks some of the true energy content of consumer nation's lifestyles.

The consumptive lifestyle, which defines affluence, consumes energy at every step. Our buildings are heated and cooled: often cooled even to uncomfortable levels in the summer. Consider all of the electronic appliances in typical affluent households: lights, multiple televisions and computers, printers, video games, washer, dryer, dishwasher, garage door openers, refrigerators (that have outgrown some of our kitchens), microwave ovens, power tools, and charging stations for an array of portable communications and entertainment devices. They all consume energy. Indeed, many of them consume energy even when not in use: requiring phantom loads to run clocks or to allow for instant start-up.

Our possessions that do not directly consume energy were made with energy. There is energy embedded in all that we use and own. Every metal implement was made from ores mined, crushed, processed with intense heat, shipped, and then processed into the finished product. The chips in our computers use abundant but very highly processed silica. Plastics and other synthetics not only require energy to make and ship, but they are themselves made from petroleum products.

The energy content of our food goes far beyond its caloric content. Modern agriculture is extremely energy intensive. Virtually all of our crops (and even more so meat) use a great deal of energy to produce. This phenomenon developed in a time when energy was one of the cheapest, most abundant commodities. It was cheaper than land or labor, so it was used excessively in order to maximize profits. Energy played a pivotal role in what many called the "Green Revolution" in which agricultural productivity soared and food became amazingly abundant and cheap for much of the world's affluent.

Adding to the energy total, much of what we consume in affluent nations is shipped great distances from where it is produced to our tables. For example, even though upstate New York is the second largest apple-growing region in the United States, it is common to see a majority of apples in upstate grocery stores from Washington State—across the continent. The island nations of Japan and the United Kingdom are extremely dependent on imported foods, as well as other products.

The more energy we consume, the more we must produce. As long as energy is abundantly available, demand levels will determine production rates for energy. As long as we demand ever more energy to do more tasks for more people, demand can only continue to grow. The most cost-effective and reliable resources available will be called on to meet the growing demand.

7.5 ENERGY SECURITY CONSIDERATIONS

Posters and bumper stickers and news reports highlight the popular perception that oil is at the heart of much international strife and war. "Hell no, we won't go for Texaco." "How many lives per gallon?" Certainly, there does seem to be a correlation between the West's willingness to go to war against a perceived despot and the presence of petroleum exports. Is the correlation direct? Do Western leaders cynically wage war in efforts to secure more oil supplies? Whether or not we accept such cynical conspiracy theory, there can be little doubt that the dependence of the Industrialized World on oil imports to keep our economies running elevates awareness of the oil producing nations over other nations.

Energy import dependence does create a vulnerability, which startled a number of affluent nations when Muslim members of OPEC issued an embargo in 1973. Daniel Yergin's massive tome, *The Prize*, speaks at considerable length about the role that oil supplies played in both World Wars (Yergin 2008). Japan's decision to attack Pearl Harbor was significantly motivated by America's sanctions on oil (and other critical) supplies. Why was Germany's most respected Field Marshall, Erwin Rommel, leading the Africa Corps? Oil. Ironically, then, it was running out of oil that sealed the fate of the Africa Corps. Fuel is especially critical to military operations, thus major energy supplies are of strategic importance. So it is natural that governments of oil importing nations are a little more aware of activities—and problems—in the exporting nations. Does this mean that the United States led a coalition into war in Iraq for oil? We would certainly hope that is not true, both for the ethical implications and because it would be utter folly to imagine that we can secure a resource for the long term in a foreign nation through invasion. Oil fields, pipeline systems, and tankers are all vulnerable to attack. Let us all hope that it was not a crass and foolish attempt to secure oil supplies.

Nevertheless, there have been and likely can be more conflicts over energy resources: especially as the global supplies become constrained. Michael T. Klare's book, *Resource Wars* delves deeply into the potential for international conflict over limited, critical natural resources as global shortages emerge. His appendix on "Territorial Disputes in Areas Containing Oil and/or Natural Gas" fills more than four pages, just enumerating contemporary sites of conflict (Klare 2001).

On top of this, an energy crisis already exists for half of humanity. Some of these people live in nations exporting vast quantities of petroleum. This breeds resentment. At the face of it, the local people see vast wealth being made and going into the pockets of a few political leaders. Increasing recognition of the role of energy in development may lead to increased resistance to exporting critical resources. A new kind of resource-based economics may emerge, in which value is tied more closely to critical resources. Much of the appeal of neo-classical economic theory is the concept that money is a universal medium of exchange. Of course, the profit-taking at each stage of a series of transactions removes some of the value.

So, when a nation exports a resource which is required to meet essential needs, the nation ultimately does without or reimports it, generally in a more refined form, at much higher cost. Value is added in the refining step, but a significant premium is charged for that value as well. Developing Nations often end up paying a large premium to import small quantities of refined product for their relatively small markets. Access to essential resources is constrained in most Developing Nations, while facing international pressure to earn foreign currency from exports, with the claim that the currency could be used to pay for what the exporting nation needs the most. The argument is that free trade allows each nation to specialize in selling what it can make the most efficiently. Unfortunately, that does not work well for essential goods. Subsistence agriculture may not, itself, support development, but it does produce food for people to subsist. Energy is likewise essential.

Resisting the monopsonistic control of the West was one of the motivations for establishing OPEC. Prior to that, the West set the prices and contracts for oil. People around the world saw oil being produced from their lands and shipped overseas. The oil generated tremendous wealth for the European and American companies and supported even greater wealth for the consuming nations. Exploitation of people's resources seemed like exploitation of the people. Perhaps not surprisingly, the first country to nationalize oil and gas was the Soviet Union, right on the heels of the Communist Revolution. Mexico and Bolivia nationalized their oil fields in the 1930s. Nationalization swept through the Middle East in the 1950s and 1960s.

OPEC was formed in 1960, but in 1973, the Arab embargo demonstrated a new level of power for OPEC. It changed the landscape of energy in many ways, although OPEC's power soon waned. Exploitation neither ended with the rise of OPEC's power, nor was it limited to westerners. The huge production of oil generated huge flows of cash through the hands of a few individuals in the ministry or central government. Vast amounts of cash attract thieves: throughout time and across cultures. So, it should come as no surprise that corruption has become rampant in some of the major oil exporting nations.

Only in a few cases is energy being used extensively to support local development. The oil fields developed by major, multi-national companies are intended for the voracious, well-established markets of the affluent nations.

Very little of the petroleum produced in oil and gas exporting nations is used to support local development. One colleague observed of Port Harcourt, Nigeria that the city was entirely dark at night—except for the light of offshore gas flares. The resentment of the local populace around these disparities leads to acts of violence and to sabotage, which exacerbates the environmental problems associated with petroleum production. Has international terrorism grown where it has because of Islamic radicals, or has Islam been radicalized in places where people observe daily the great disparities in wealth and power supported by the vast wealth of exported oil? Terrorism and unrest are, of course, not restricted to Islamic locales, but occur wherever people see enormous disparities in wealth and power.

Gas flares, which are symbolic of the disparities in places like Nigeria, rightfully draw attention. Gas is much harder to export than crude oil. Gas is convenient to transport and to use locally, but for trans-oceanic movement, it must be liquefied to extremely low temperatures (−259°F or −161°C) and transported in special LNG tankers. The required infrastructure adds well over a billion dollars to the cost of developing a gas field. Consequently, oil remains the preferred commodity for export, which is almost always the target for major companies. The local markets are small, with the world's poor consuming about 1% as much energy per capita as affluent people; the currency is typically unstable in developing countries. So, gas, extremely valuable for local use, is often deliberately by-passed or flared as a waste product in developing countries, where it is needed the most.[1]

The focus on energy for the affluent world leads to ignoring important potential to support development. Many nations are recognizing the value of energy for local development, evidenced by the increasing frequency of production sharing language in international oil and gas contracts. When oil was discovered in Uganda in the early 2000s, the Ugandan president made a public statement that he did not want any unrefined crude oil to leave the continent. This reflects the value of energy to local economies, including the value of refinery operations and the jobs associated with that phase of the business.

There may be increasing pressure by governments in developing countries to keep energy products within their border, to support local development. It can be argued that the currency earned by exporting petroleum is much more fungible, but that very fluidity is what leads to

[1] This was a large part of the impetus for the nonprofit AHEAD Energy, with which both of your authors are involved.

corruption. Furthermore, it should be apparent that energy, to the extent it is needed, has greater value than its wellhead price. Otherwise, no one would be willing to pay for the crude oil, only to pay even more to ship it to consumers overseas.

7.6 COMPARING THE VALUES OF DIFFERENT ENERGY SYSTEMS

7.6.1 FOSSIL FUELS

The fossil fuels are notorious for their environmental externalities and their stock-limitations. Some authors and researchers like to dismiss them as "obsolete." Why, then, do they still dominate the world's energy markets? The answer is that they have several benefits that are difficult to match. One is the concentrating and processing that nature has performed on biomass over millions of years to convert it into rich stockpiles of fossil fuels. Author Paul Roberts notes, in Mother Jones magazine, that "Oil's qualities were unbeatable when it cost $25 per barrel and even at $100 per barrel… ." He notes not only the natural processing, but the fact that the fossil fuels are isolated from the larger economy. Some of the alternatives (such as ethanol from corn), are tightly linked to the larger economy in that they take land and water that could also be used to produce food (Roberts 2008).

For all of the problems associated with them, fossil fuels offer millions of years of accumulation, compaction, intensification, thermal and chemical processing, all of which results in high value fuels. Processing of biofuels, to some extent, seeks to emulate these fuel values. So, we should appreciate the high quality fuel sources that nature has accumulated for us over the years—but recognize the environmental and social impacts of extraction, along with the limitations of these fossil fuel resources that took millions of years to accumulate.

The millions of years of accumulation and natural processing resulted in very rich, energy dense fuels. Their value keeps the fossil fuels central to the economies that can afford to produce and/or to purchase vast quantities. The alternatives face a steep challenge in competing with the concentration and energy and power densities of the fossil fuels: even their cleanliness is superior to less refined, raw biomass. Nevertheless, each fossil fuel has very real limits in relation to sustainability.

Coal is the world's most abundant fossil fuel resource, yet one of the most problematic. High GHG emissions and the disruptive nature of coal mining make it less than ideal as an energy provider for the near future. Improved coal boiler technology can mitigate some of these problems, but increasingly severe environmental problems, such as acid rain and soil contamination, coupled with the fact that coal resources will ultimately be depleted as well, will still result in a relentless decrease in resource value over time.

While petroleum fuels burn cleaner and have higher energy contents than coal, the release of GHG and the fact that the supply is ultimately limited will result in increasingly severe costs over time. The fuel value lost through excessive consumption, resulting from low petroleum prices will be especially apparent once a shortage hits, as there is really no viable alternative for gasoline and diesel fuel for use on the scale it will be needed in the near future.

Natural gas avoids many of the major external environmental costs associated with the other fossil fuels, yet its paradoxically low price, and finite supply still dictate that there are limits to the value gained in the continued expansion of this fuel source. It has clearly greater benefits for local consumers than in the global energy markets.

At the International Solar Energy Experts workshop held at Kwame N'krumah University of Science and Technology in Ghana in late May, 2012, over a lunch discussion, colleagues asked, "do you believe fossil fuels are better than renewables?" No, but they are more mature. It's like the reality that we older people will reach the end of our productive lives and be replaced by a new generation. The new alternatives will offer some improvements over the older generation of fuels they replace, but the global society is not ready to forego the contributions of the older generation and turn our needs over to the callow, future generation of energy systems.

7.6.2 RENEWABLE RESOURCES

Renewable resources have been heralded as the white knights of energy supply, yet a complete life cycle analysis does show areas where these resources fall short of expectation. It is important when researching any system to define the scope and boundaries appropriately, which has not been the case in many studies on renewable resource use. Ethanol to replace petroleum is a classic example. After shuffling energy from one source to another, it would appear that ethanol offers little or no net energy gain, especially if derived from corn grown as a fuel crop. Again, we believe that it will be more valuable to evaluate biomass-derived energy as a stock-based and flow-based combustion resource. Human use taps into reserves that can be depleted, its renewal rate can be exceeded by the rate of consumption. It should have assigned realistic costs assigned to account for air pollutants, as well as water and land use.

The conversion of solar energy to usable energy in the form of plant life is clearly of value. The production of ethanol from corn, though, incurs the costs of energy intensive processes relying on depletable fossil fuels. There are two primary motivations for displacing gasoline with ethanol: reducing CO_2 emissions to the atmosphere and reducing dependence on a stock-based, depletable, (and largely imported) energy source. They are both worthwhile goals. To the extent that the value of ethanol is predicated on either of these benefits, though, we must be clear the extent to which it actually yields a true net value.

For example, studies have reported that there is a net energy gain when converting corn to ethanol in the range of 20% (Wang et al. 1999; Wang et al. 2007). Even ignoring studies with contrary results, one question that needs to be asked is whether a 20% energy gain is worth the investment of time, money and other resources necessary to create it. The cropland converted to produce this corn was once a contributing part of a much larger ecosystem. The resources invested in obtaining fuel to power our cars are diverted from growing crops for food. Land and water use are additional externalities that need to be assigned to producing ethanol from corn.

Even if biomass production is tightly regulated to constrain consumption to renewable levels, there will be negative health effects associated with the use of biomass in closed quarters. Raw biomass burns less cleanly than more dense fuel sources, and while carbon emissions can be considered negligible by citing a closed carbon cycle for well-managed growth, smoke and particulates still present major health concerns. Most often, however, biomass is not used at a sustainable rate, resulting in localized deforestation and public safety concerns that all detract from the value of the resource. Waste derived fuels avoid the energy, land use, and environmental impacts of crops grown for fuel.

Capturing solar energy using PV cell technology could provide a practically unlimited supply of energy for consumption. There are economic barriers to tapping solar energy, but also external costs that decrease its value. PV cells themselves use exotic materials which may have economic constraints and environmental impacts, and which will need to be accounted for in some manner. The need for storage also detracts from value.

Wind is a practically unlimited source of energy where available. It is constrained by issues of scale. Since harvestable wind power increases with the cube of wind speed, tall towers, reaching unobstructed winds are able to capture much more energy than those nearer the ground. This is an added factor favoring large-scale development. The scale incurs "not in my back yard" concerns, as people often do not favor the sight of 300-foot-tall wind turbines dotting the landscape. This, in turn, favors remote placement of the large-scale wind farms. The need for storing energy where wind is intermittent adds incentive to remote citing, where high wind energy is constantly available. Remoteness begs questions of efficiency losses in long-distance electricity transmission.

Geothermal reservoirs can be depleted in the medium term, even though the overall resource is not being depleted. Heat flows inside Earth will ultimately replenish depleted reservoirs, but over what time? Earth's heat is ubiquitous, but current technologies only tap relatively rare occurrences of natural steam reservoirs in rock that is porous permeable (often due to natural fractures). Innovation may enable tapping far more of the hot rocks that are relatively shallow—through extensive induced fractures—perhaps like those being hotly contested in the Marcellus shale, to extract natural gas. Will it be feasible to tap the heat from lower enthalpy (cooler) reservoirs? At depth, the rocks and fluids contained therein are always hotter than the surface. Since geothermal has the characteristics of both stock and flow limited resources, it does not require storage.

Hydropower has been the most successful of the noncombustion energy sources, historically. It clearly has no atmospheric emissions (except those incurred in building the facility) and taps a relatively nondepletable source. The ecosystem and sediment transport disruptions must be assigned reasonable costs as externalities, though, to arrive at realistic values. There is also often a social cost in displacing people who live on land that will be flooded by the reservoir. Some projects, such as the Niagara Falls hydropower system, avoid such disruptions. In fact, the water diverted from the Falls, actually slows the erosion that will ultimately eat the famous falls away. Such applications limited to tapping a natural flow at a falls are quite rare.

Small-scale hydropower has the potential to avoid many of the externalities of large-scale hydropower. Indeed, work on ever-smaller systems has progressed through: mini, micro, and now pico hydropower. This trend fits well with the proposed trend toward more *decentralized* power systems. Unfortunately, while having minimal externalities, the direct cost tends to be high, since these projects fail to benefit from economies of scale. In the Developing World, though, where more energy production is so desperately needed, the opportunity to realize incremental improvements for small capital investments may offset the problem of higher costs per installed watt.

7.6.3 NUCLEAR POWER

The value to be realized by nuclear energy is highly speculative, as there is no consensus on what the risks associated with nuclear power generation are. Obviously, there is a negative value associated with the waste products produced by nuclear power plants, but the cost and scope of disposing of these byproducts is under debate, as are the risks of expanding nuclear power generation in the near future. For these reasons, nuclear energy's potential value will be limited by the limited social acceptance of increased nuclear power. We believe that a robust metric for valuing energy will acknowledge social acceptance criteria. Even fears that are not entirely rational cannot be discounted when assessing value. Certainly, safety concerns have limited the growth of nuclear power over the last 40 years. We think that they will continue to limit its actual growth, at least until global shortages raise prices sufficiently to compensate for the concerns. Thus, the near term potential value added by nuclear power will likely be small, similar to the technically immature technologies.

7.6.4 HYDROGEN AND FUEL CELLS

You may have noticed that this text has said relatively little about hydrogen fuel cells, which some have touted as the answer to sustainable energy. One reason that we have not discussed hydrogen fuel cells extensively is that this text is focused more on energy sources and consumption than on intermediate technologies. Hydrogen is not an energy source on this planet. It does not exist on Earth in its elemental state, so we must expend energy that we control to produce hydrogen—then to store it. The energy input is definitely greater than the output, so hydrogen is not an energy *source*, it is a *carrier.* It is like electricity or gasoline, or a battery.[2] It can hold and transport energy, but it must be derived from an original source.

Advocates of hydrogen fuel cells point out that a fuel cell, powering an electric motor can be much more efficient than a gasoline internal combustion engine. This is true. The overall energy efficiency and environmental impact are functions of how the energy is generated initially to split hydrogen from other compounds. Furthermore, what is the source of most hydrogen being marketed today? Natural gas. Mixed with high temperature steam, hydrogen splits off from methane (one carbon atom attached to four hydrogen atoms), producing hydrogen and carbon dioxide. This can be done at relatively high efficiencies of about 80%. Nevertheless, it still relies on a depletable resource as the primary feedstock. Natural gas can also be burned directly for heat at extremely high efficiencies (greater than 95%) or to generate electricity at relatively high efficiencies. This is clearly not an alternative source.

The dream of hydrogen fuel cells is to split it from water, or to generate it biologically, with either method using a renewable energy source, like solar or wind. Some, though, raise questions about what is gained by going through all that is required for hydrogen fuel cells. The hydrogen must be generated, then stored either at extremely high pressures or low temperatures (all of which takes energy.) The fuel cell car being developed by several automobile companies has anticipated storing compressed hydrogen at some 10,000 pounds per square inch—that is more than 700 times atmospheric pressure. Then the fuel cells are expensive. Certainly, the price can come down if the fuel cells go into mass production. For vehicular applications, there is not only the hazard of storing gas at extremely high pressures, but the dramatic cycling between high and low production rates that attend acceleration and deceleration tend to wreak havoc on the membranes that are the heart of hydrogen fuel cells.

---

[2] We have not delved into the technologies of refining crude oil into gasoline, or of generating electricity in power plants, or of batteries either.

Fuel cells may become highly useful for stationary energy storage, but the problems seem serious for the vehicular applications that have drawn so much attention. Stationary applications can use a variety of kinds of fuel cells, some of which use fuels other than hydrogen. Stationary applications can also produce at more stable rates. Fuel cell technologies may contribute to more efficient and cleaner technologies, but they are not yet likely to transform energy systems radically. Despite this, we do acknowledge the appeal of a hydrogen economy; however, right now it should be viewed as having prohibitive obstacles.

7.7 EXTERNALITIES IN ENERGY VALUE METRICS

As we close this chapter, we would like to emphasize that full systems-level analyses should be applied to all energy alternatives—including those currently in use. More sustainable energy systems enhance the net value derived from energy, just as sustainability, in general, supports development today without sacrificing potential for future generations. Contexts must be considered in these analyses. The needs, challenges, and opportunities are all quite different for different people in different times.

Additionally, societal factors are not well captured in most of the metrics used to compare energy sources. We consider the failure to account for social factors to be critical. Moreover, practically no existing metrics include any explicit time value to account for the issues of depletion (reducing future value for a system based on depletable energy sources) or for technical maturity (limiting current value.)

The process of evaluating the value of energy will not be universally agreed upon, nor does it have to be. There will be assumptions inherent with every approach. An effective metric must capture the range of costs and benefits of each energy alternative. The externalities—the costs, or benefits, not included in the market price, must be accounted for. But how to assign values that the marketplace does not? At a talk presented at the University of Rochester, Herman Daley spoke of the need to internalize externalities.[3] A student in the audience challenged him, arguing that it was impossible to assign accurate values to externalities. Daley replied to the effect, "I don't care what value you assign to a negative externality, just so long as it's not zero." He elaborated that it would not be unreasonable to set externality values **deliberately low**; the important first step was to acknowledge that they exist and begin to incorporate their costs into economic analyses.

There are environmental and resource costs associated with every human activity. Environmental issues are sometimes conflated with aesthetic and other personal values. As we attempt to suggest means to compare energy options on relatively equal footing, it is imperative to arrive at relatively universal understandings of the environmental impacts or externalities of each energy system. In this effort, we suggest deconvolving intrinsically subjective values (e.g. aesthetic or moral) from those which have objective factors.

Energy value modeling cannot be perfect, but we suggest that steps need to be taken to make more robust models, which more closely encompass all of the costs and benefits of energy systems on even footings. It will be a dynamic process, rather than a singular decision. That being said, there are data available to assist in creating a framework for assigning energy values,

[3] Daley, an eminent economist, co-authored the book *For the Common Good* (1994) with eminent theologian John Cobb—an interesting partnership and a seminal book on sustainability.

and some reasonable assumptions upon which to model future developments. Ideally, energy value modeling will be an ongoing effort to include improved understanding of the externalities and to accommodate changing conditions.

Variations in methodology and scope of work have caused past energy externality studies to differ greatly in their findings (see Sundqvist 2006). Even so, certain trends are apparent throughout. It can be confidently stated that fossil fuels have higher external costs associated with them than some of the renewable or flux-limited resources, notably: wind, solar, and run-of-the-river hydropower. Nevertheless, it is folly to assume that any energy resource will be without negative externalities as it begins to produce at ever-larger scales. There will most likely be externalities that emerge as the alternatives are produced at much larger scales. This is a topic that will need to be revisited regularly.

There are a number of factors to consider when determining the external cost of energy. Legislation already seeks to address some of the more conspicuous externalities, while others remain to be acknowledged. In many low-income countries, fuelwood and diesel fuel sell for similar prices. Reflected in these prices are the costs of acquiring the fuel, transportation, and distributing it to the consumer. While similar in price per energy content, the diesel fuel (or kerosene) has health, safety, and practical utility advantages that make it a more valuable resource. If it could be assumed that the fuelwood was only being marketed at a sustainable rate (something rarely done in rural Africa) then it may make up for its health deficit in value added because of resource sustainability—but this is not the case. External costs may never be seen by the consumer, but the debts incurred by them will eventually need to be paid: most likely with shortened lives and endangered ecosystems.

A starting point for each of these energy value considerations is to characterize the quantity of each fuel source being used, its established rate of growth over time, and its potential. The exponential growth in demand is the largest single factor in limiting the time horizon for stock-limited resources, such as petroleum. In undertaking the peak analyses discussed in Chapter 5, the exponent used (1% per year rather than 2% per year) has more effect than assumptions about the petroleum resource base. Facilitating growing consumption by maintaining low prices will only magnify the problems as they begin to arise.

Relatively little work has been done to assess the value impact of depletability or other forms of resource limits. Some simply discount the depletable resources, with epithets (or epitaphs) such as "obsolete petroleum" or "oil is the resource of the past." These dismissals, we suggest are misleading and counterproductive. If, indeed, we saw nondepletable alternatives taking noticeable market shares from petroleum, or constraining demand for it, then these dismissive remarks might be appropriate. Such is not the case. Petroleum demand has continued to rise through the first decade of the new millennium, abated only by the "Great Recession" that started in 2008, which also suppressed demand for many other commodities. The reality is that petroleum continues to hold the lion's share of global petroleum markets. Stock-limited resources add value. We must, however, find a means to capture their diminishing value with depletion. We suggest that this task requires a metric incorporating a time function.

The challenge will be to develop robust metrics that assign appropriate costs to externalities including depletion, safety, and a full range of environmental impacts, while assigning appropriate values based on the potential of each resource. The impacts must include those incurred in acquisition, processing, storing, transporting, using, and disposing of the products.

REFERENCES

Blok, K. 2008. *Introduction to Energy Analysis*. Amsterdam: Techne Press.

Daly, H. E., and J. B. Cobb. 1994. *For the Common Good: Redirecting the Economy Toward Community, the Environment, and a Sustainable Future*. Boston: Beacon Press.

Ebenhack, M. J., and B. W. Ebenhack. 2008. "A Role for SPE in Meeting Vital Human Needs." 9th International SPE Conference on Health, Safety and Environment, Nice, France.

FAO (Food and Agriculture Organization of the United Nations). 2005. "Incentives to curb deforestation needed to counter climate change." http://www.fao.org/newsroom/en/news/2005/1000176/index.html

Goldemberg, J. 2001. "Energy and human well-being." United Nations Development Programme Report. New York: UNDP.

IEA (International Energy Agency). 2002. *World Energy Outlook 2002*. http://www. worldenergyoutlook.org/.

Klare, M. T. 2001. *Resource Wars: The New Landscape of Global Conflict*, New York: Henry Holt and Co.

Lim, S. S., T. Vos, A. D. Flaxman, G. Danaei, K. Shibuya, H. Adair-Rohani, ... and L. Degenhardt 2013. A comparative risk assessment of burden of disease and injury attributable to 67 risk factors and risk factor clusters in 21 regions, 1990–2010: a systematic analysis for the Global Burden of Disease Study 2010. *The Lancet*. *380* (9859): 2224–2260. doi: http://dx.doi.org/10.1016/S0140-6736(12)61766-8

Martínez, D. M., and B. W. Ebenhack. 2008. Understanding the role of energy consumption in human development through the use of saturation phenomena. *Energy Policy*. *36* (4): 1430–1435. doi: http://dx.doi.org 10.1016/j.enpol.2007.12.016. Figure 7.5 reprinted with permission from Elsevier. Copyright (2013)

Pasternak, A. 2000. Global energy futures and human development: a framework for analysis. US Department of Energy Report UCRL-ID- 140773, Lawrence Livermore National Laboratory. https://e-reports-ext.llnl.gov/pdf/239193.pdf

Rajagopal, D., and D. Zilberman. 2007. *Review of Environmental, Economic and Policy Aspects of Biofuels*. World Bank Report WPS4341. doi: http://dx.doi.org/10.1596/1813-9450-4341

Ramanathan, V., and G. Carmichael. 2008. Global and regional climate changes due to black carbon. *Nature Geoscience*. *1* (4): 221–227. doi: http://dx.doi.org/10.1038/ngeo156

Roberts, P. 2008. "The Seven Myths of Energy Independence." Mother Jones, May/June. http://www.motherjones.com/politics/2008/05/seven-myths-energy-independence

Smil, V. 2003. *Energy at the Crossroads: Global Perspectives and Uncertainties*. Cambridge: MIT Press.

Smith, K. R., and R. Cowlin. 2005. *Household Fuels and Ill-Health in Developing Countries: What improvements can be brought by LP Gas?* World LP Gas Association and Intermediate Technology Development Group (Practical Action), Paris. http://ehs.sph.berkeley.edu/krsmith/?p=335

Sundqvist, T. 2004. What causes the disparity of electricity externality estimates? *Energy Policy*. *32* (15): 1753–1766. doi: http://dx.doi.org/10.1016/S0301-4215(03)00165-4

UNDP (United Nations Development Program). 2005., *Energizing the Millennium Development Goals*. New York: UNDP.

Wang, M., Saricks C., and D. Santini. 1999. "Effects of Fuel Ethanol Use on Fuel-Cycle Energy and Greenhouse Gas Emissions." Center for Transportation Research, Energy Systems Division, Argonne National Laboratory.

Wang, M., M. Wu, and H. Huo. 2007. Life-cycle energy and greenhouse gas emission impacts of different corn ethanol plant types. *Environmental Research Letters* 2. doi: http://dx.doi.org/10.1088/1748-9326/2/2/024001

Wines, M. 2007. "Toiling in the Dark." *International Herald Tribune*, July 28, 2007.

WRI (World Resources Institute). World Resources 2008: Roots of Resilience—Growing the Wealth of the Poor. Washington, DC: World Resources Institute. http://www.wri.org/publication/world-resources-2008-roots-of-resilience

Yergin, D. (2008). *The Prize: The Epic Quest for Oil, Money & Power*. New York: Simon and Schuster.

CHAPTER 8

Next Steps

In this last chapter, we discuss possible and probable scenarios for energy use that will be encountered in this century.

8.1 ENTERING A NEW AGE

The path to sustainable energy involves exploring new territory. We have certainly not been there before, and it will be much more like the journey of Lewis and Clark than of the pioneers on the Oregon Trail. We know a general direction and some better and worse headings in which we should set off, but the trail has yet to be blazed. Yet, this journey is inevitable. Like entering adolescence, it is a rite of passage about which we have no choice. There will doubtless be a pervasive theme of trauma and transformation. We can only choose how we will proceed: learning and maturing or clinging pathetically to what was or to what we wish, rather than to what can be. The practical realities of what is possible and in what time frame will be disturbing to many of us.

Sheik Ahmed Zaki Yamani is oft quoted as saying, "the Stone Age did not end for lack of stones and the petroleum age will not end for lack of petroleum." While Sheik Yamani was probably both a shrewd and thoughtful leader who guided OPEC from obscurity into an influential role in the world, this particular quote is misleading. One may even question whether he himself believed it or found it the diplomatically appropriate thing to say at a time when anxieties were running high in the Industrialized World over what many found to be the newly discovered possibility of running out of oil.

At any rate, Yamani's analogy is flawed. The Stone Age did not come about due to the extraordinary merits of stones. Stones were readily available to pick up and use. Some of them could be shaped, but their use was largely opportunistic. They had only to compete with bare hands and sticks in their rise as tools. Eventually, the Stone Age gave way to the superior qualities that refined metals offered in the Bronze Age. The rise of the Petroleum Age, however, was very different. Petroleum out competed other energy sources, quickly gaining a remarkable market share in just a few decades. It subsequently gained a lion's share once it was used widely for transportation and for generating electricity (primarily from gas), and has held it in spite of millions of researcher hours and billions of dollars of government subsidies to try to supplant it. It has clear and present advantages over its competition, unlike stones. The alternatives are known, but have yet to be able to compete. Petroleum is not a primitive tool waiting to be displaced by far more powerful or versatile alternatives.

Thus, as we consider the steps toward coming energy transitions, it is worthwhile taking a look at the original transition to petroleum and the likely prominent role it will continue to have going forward.

8.1.1 THE TRANSITION THAT BROUGHT US HERE

As discussed in Chapter 2, petroleum's original market niche was to compete against increasingly scarce whale oil. In modern times we don't really think about whales as a resource, rather focusing our attention on the ethical issues of exploiting and killing relatively sentient creatures. The 19th-century problem, though, was clearly one of the limits to renewability of a biofuel source. Intensive hunting was impacting the whale species and making it increasingly more difficult to harvest whale oil. Laboratory tests had already demonstrated that the distillation product of crude oil known as kerosene had superior lighting properties and a cleaner burn than whale oil. Even the meager production of Drake's famous discovery well dwarfed whale oil operations. Consider that a successful whaling ship, such as the Charles W Morgan, would average a production rate of less than two barrels of oil per day—the 10 barrels of oil per day from a single well drilled by a handful of men was astounding (Mystic Seaport 2012).

Soon single wells were being drilled that produced more crude oil than entire fleets of whaling ships. The product was also of superior quality—kerosene burned cleaner and brighter than whale oil. The versatility of the product was also tremendous, within a few decades adding gasoline to the list of petroleum products, as well as lubricants and asphalt products. Then came petrochemicals and the production of synthetics, which have revolutionized our world in so many ways. Most reading this book are wearing petroleum products, sitting on them, drinking from cups or bottles made from petroleum, and even ingesting them in the form of pharmaceuticals and food additives partially derived from petroleum products.

Refinery design can be changed to account for shifting demand profiles for the wide range of petroleum products. The earliest refineries sought to maximize kerosene production, whereas the rapid growth in demand for gasoline soon prompted refineries to shift their technologies toward maximizing the output of that product: especially in the United States, where diesel had relatively small market shares. Petroleum (oil and gas collectively) came to dominate global energy markets because they were superior energy sources that could be produced in vast quantities, were highly portable, enabled the development of fast powerful engines, burned cleanly, and most of all were cheap.

Petroleum has become thoroughly entrenched in Western society by virtue of the benefits that it has offered. Any alternative energy source will have to compete against these benefits, including low-cost, as long as oil and gas are sufficiently abundant. In affluent economies, alternatives are also likely to have to compete against the established infrastructure for petroleum consumption: pipelines, refineries, gasoline stations. These infrastructural components are already in place, and their costs have been fully amortized. New systems that do not utilize the same infrastructure, incur additional costs to develop new infrastructure

Consider the United States transition to petroleum and the improbable transition away from it in Figure 8.1. This pair of images illustrates a challenge. Petroleum is capable of producing

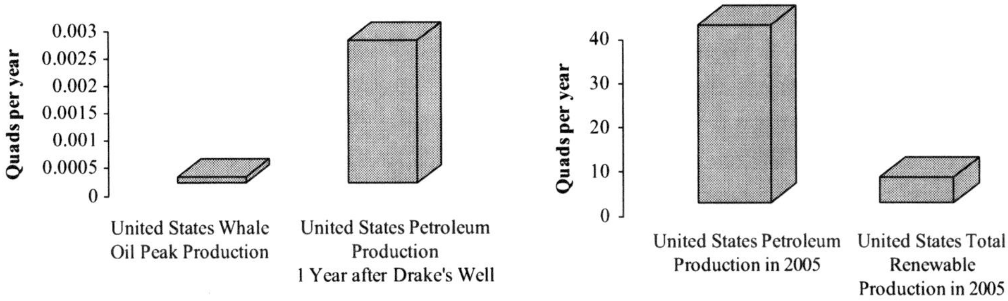

Figure 8.1. A comparison of the transition to and from petroleum.
Source: Ebenhack (2010).

vast quantities of energy, with relatively little input. Whereas it was known and used marginally for centuries, the industry that found that it could flow from wells flourished almost overnight: quickly dwarfing its primary competitor, whale oil. It was a cleaner, more efficient illuminant, which was more easily acquired in tremendously greater quantities. The same cannot be said for the alternatives being proposed to displace petroleum. Decades after major research and development were launched, they still represent miniscule production, compared with petroleum. Some of the alternatives will succeed in taking over some of petroleum's market share. They simply are hard-pressed to compete as long as petroleum is able to provide a supply that exceeds demand. It is true that wind and solar photovoltaic (PV) energy production have grown dramatically of late, but consumption was growing too until the most recent global recession.

8.2 PETROLEUM'S ROLE IN THE NEXT TRANSITION

At a recent conference of the International Society for Engineering Education in Cork, Ireland, one speaker related a popular Irish anecdote that a traveler asks someone on the roadside how to get to Dublin. The response is, "Well, I wouldn't start from here." Just so, for energy transitions, we wouldn't start from here. But we must. Fossil fuels, and petroleum in particular, are the starting point and we must recognize this in order to plan an effective transition path.

In light of the value that petroleum and its products have brought and continue to bring to consumers, its demand will not be easily displaced. Since we believe that we have offered strong evidence in Chapter 5 that the petroleum decline will not be abrupt, oil and gas collectively are likely to remain important energy providers into the 22nd century. They will probably fall short, however, of keeping pace with growing demand before the middle of the 21st century. Though, the latter prediction seems the more reliable of the two.

As long as petroleum is able to satisfy marketplace demands, the alternatives face a near insurmountable challenge to take over significant market share. Not only are petroleum products extremely energy dense, versatile energy carriers, but they have set a price standard that is too low for alternatives to compete effectively. The good news for alternatives is that the low price cannot and will not continue indefinitely. Prices for oil and gas will rise as the global production peak is approached. So, the alternatives will not need to compete with current price structures for very long. It is a fallacy to evaluate the potential of alternatives in comparison to current energy prices, established by the abundance of oil and gas.

The great challenge for alternative energy systems will be to achieve significant levels of production when they are needed. Consider, for example, a comparison of "renewable" energy production to petroleum's surplus production capacity. Since energy is an essential commodity, the marketplace requires surplus capacity to function normally. That surplus, unused, production capacity for oil has tended to run between two to four million barrels per day over the last couple of decades: about 5% of the global total production. That surplus is more than the energy equivalent of all of the global solar, wind and modern biofuels production combined. This means that the popular alternatives can barely offset the minimum productive capacity that must be held in reserve for petroleum to maintain stable markets. (It may be noted that the surplus production capacity dropped during the period 2003–2008. During that time, prices rose dramatically.)

As we do approach the peak and global supply shortages emerge, though, the price can be expected to increase and remain much higher. If hypothetical demand continues to grow, and alternatives are not abundantly ready, prices would be expected to rise until they push sufficient demand out of the marketplace and/or draw enough alternative sources into the marketplace to restore not merely a balance, but a sufficient production surplus capacity. Again, since energy serves essential human needs, a functioning marketplace demands that it is available in surplus. Consider having just barely enough food, or air, or water to survive. An asthma sufferer (like Ebenhack) can assure you that a bare minimum of air would not seem at all sufficient. So, energy also needs to operate with a surplus. As soon as production capacity can only equal demand, the system is operating in market failure conditions (Decanio 2006). Therefore it is very important for society to plan for the coming energy transition and to do so with the best information possible. In a market failure, prices will continue to rise, but will not succeed in bringing production and consumption into a prompt balance. Of course it is impossible to consume more than is produced, but there will be real shortages in the marketplace in the times of market failure.

The market failure will be neither an absolute nor a permanent phenomenon. Rising prices will provide companies with the incentive to find and produce erstwhile marginal resources. This is significant because it makes understanding the reality of an impending shortage far less dependent on assumptions about how much oil remains to be discovered in the largely unexplored provinces of Africa, Central South America, Central Asia, and the Arctic. It reduces dependence on assumptions about the future significance of shale plays, enhanced oil recovery (EOR) or exploration technologies. These factors will undoubtedly continue to increase our perception of how much oil can be recovered from Earth's crust—especially after rising prices increase the incentive for more aggressive exploration and implementation of new technologies. The rising prices will follow the onset of the crisis. Therefore, much of this additional oil will serve to make the decline more gradual, not to forestall the shortage significantly. This is made even clearer by understanding that the crisis will begin before the peak is reached.

But there is some good news. Rather than the problem, rising prices are part of the solution. They will squeeze some of the superfluous, discretionary consumption out of the marketplace rather quickly. They will support the deployment of more efficient technologies and suppress the least efficient. (For example there will be an increased demand for efficient cars and very little demand for "gas guzzlers.") Energy will be less taken for granted, which may draw more of the best and most thoughtful talent to the problem of sustainable energy systems planning. Finally, higher energy prices will support the alternatives. Those that are ready to compete will gain market share as quickly as the production technologies and infrastructure can grow.

8.2.1 PETROLEUM'S RESPONSE TO THE SHORTAGE

The economic constraints will change as prices rise. Technologies that cannot compete with cheap oil will enter the marketplace (and yes, it's still cheap at three dollars or even four dollars per gallon in the United States. Even six to eight Euros per gallon in Europe will likely be found not to be a high price for refined petroleum products.) Resources that have been overlooked will draw renewed attention. Through much of recent history in the oil business, there has been about a two order-of-magnitude disparity between the economic thresholds for domestic (United States and Western Europe) oil field development versus international projects. That is to say, if we drill a moderate depth 5,000 foot (1,500 meter) well outside Casper, Wyoming and it produces 50 barrels of oil per day, it's likely to be a commercial discovery. A well outside Vilanculo, Moçambique would need to produce the equivalent of 5,000 barrels per day—or more—to be commercially attractive for export.[1] Literally thousands of oil and gas wells have been drilled around the world that have encountered oil and gas, but not enough to be developed for the export market. The data from these "sub-commercial" wells will merit new attention when prices rise significantly.

Then there will be "reexploration" to develop some of the oil that has been discovered, but never produced. If an exploratory well finds oil (or gas), but not in sufficient quantities to justify development at prevailing oil prices, it is abandoned without ever producing a drop. (As mentioned above, this may be particularly true in lower income nations, but can also be true in any challenging location.) Remember that global oil prices were in the teens through the 1980s and exceeded 100 dollars per barrel in 2008 and again in 2010. A reservoir that was sub-commercial at 15 dollars per barrel may be well worth producing even at less than six times that price. A good deal of secondary exploration will take place in file rooms, reviewing the logs and well tests from what had been deemed "dry holes" under a different set of economics.

Reexploration will bring attention back to many regions where exploration efforts failed to find commercial quantities at lower prices. So, more of the developing nations will have the opportunity to become energy exporters. They will need to consider carefully the value of the revenues versus the value of the energy pumped into their local economies. There will be an altogether different playing field for energy once the shortage hits. There is some evidence suggesting that energy consumption is more valuable to a society than the monetary gains from selling the product on the export market (Khodeli 2007). This makes sense in that importers would just keep the money if it were intrinsically more valuable than the energy. (Of course, if a country can produce energy beyond its needs, exporting some of the surplus can bring in currency to buy other things.) We have also seen some evidence that the huge cash flow through a few hands often does not make it to the majority of the people effectively, for a variety of reasons (Humphreys et al. 2007). So, as the crucial role of energy in development is better understood, developing nations may be less willing to part with their energy wealth. The obvious response on the part of importing nations is to bid up the price on energy. An equilibrium will be reached, but probably at a higher price and, perhaps, with a larger share of the petroleum products remaining in the producing nations, to support development.

Furthermore, as the prices rise, the recovery factors for existing fields will increase. Since conventional production techniques barely produce one-third of the original oil-in-place from any given field, there is considerable potential to expand the ultimate recoverable reserves

[1] Recently Anadarko Petroleum Company made a giant gas discovery off the coast of Moçambique, a country long ignored by international exploration.

base—perhaps more than double it with EOR. However, the real potential of EOR is not likely to be realized until prices remain consistently high for a few years—which will almost certainly be after, or at least during, a global production peak. It is for this reason that we see a peak occurring not much later than the gloomy peak oil acolytes do, while we say that the world may ultimately yield up to three times as much oil as they believe. Much of it will be after the peak.

8.2.2 THE TIME FACTOR

It is high prices that will spur much of the new and unconventional exploration, reexploration, and EOR. All of these factors will come into play primarily after the peak and subsequent high energy prices. More aggressive drilling and technologic innovation in response to higher prices that come from global shortages will not likely provide the production capacity to meet growing energy demand for much longer, but will ease the transition somewhat. We believe that's the best news we can give you. Petroleum will be around for a long time, providing significant amounts of energy, but not promoting unlimited growth.

Recall that Wolfgang Schollnberger (2006) has described a number of characters possible for the peak to assume, from relatively sharp, to a long plateau, or long, gradual decline. We are inclined to believe that there will be a long plateau, but suggest that, if viewed in detail, it is likely to be quite irregular. Part of this is due to uneven response times. Higher prices can have a short-term effect of depressing demand (especially in the United States where so much of our consumption is discretionary: people can decide that they won't take that road trip to Yellowstone this summer.) These reductions in demand will cause near-term price decreases, but much of the deferred demand will return as soon as prices dip. Sustained conservation efforts are likely to take time. People don't instantly go out to buy more efficient cars (and appliances, and certainly not homes.) It also takes time for the shifting consumer preferences to filter back into new product design. Even in times of four dollars per gallon gasoline prices, consumer articles evaluating "high mpg" vehicles tend to put a priority on "peppiness" and other such discretionary qualities, which are innately energy consumptive. It will take real time to displace consumer preferences based on perceptions of power and speed, with a preference centered on efficient transportation. People need to decide "I don't seek to imagine myself as a racecar driver or lumberjack or cowboy, but rather seek to move myself and my family efficiently and safely from one point to another." It's a different mindset. Once automakers perceive a change in mindset, they will rush to make more efficient vehicles. But, even once they start the rush it will be a few years before the new, efficient cars make serious inroads in transforming the fleet.

In the meantime, the overall trend will be one of rising prices. These higher prices will promote greater efforts to find and produce more oil. But, again, there will be a lag time. Successful EOR projects will take perhaps five years to begin boosting production in the best fields. If the global shortage is perceived clearly, then companies will embark on many such projects at once, along with more aggressive exploration. This may result in a new mini peak in global production. But the EOR projects will only yield modest increases in production rates; they will commonly do more to slow decline than to increase production rates. The more aggressive exploration efforts will include completing wells that would have been deemed sub-commercial before. Each of these will add very little to global production. Many of these wells, producing from small reservoirs, will deplete rather rapidly.

So, we can expect some modest "ups and downs" at the peak time, but the overall trend will be one of slow production decline, with relatively higher prices. This will characterize the peak and postpeak era until alternatives can effectively take over significant market shares from petroleum. As the alternatives sort out winners and losers in open competition, some will probably achieve new production efficiencies, which may ultimately usher in a new era of cheap, abundant energy. Someday, but not soon.

The irregular character of the peak time is actually likely to make it hard for the alternatives. Higher prices will help, but the occasional dips in prices that accompany occasions when conservation and increased production coincide could be devastating to start-up alternative energy companies. Major oil companies weather the vicissitudes of market instability in part because they have enough assets to carry them through, but we've even seen many of the major oil companies struggle and end up merging with or being bought outright by other larger companies. Many small oil and gas companies go out of business every year. The same is likely to be true of many alternative energy enterprises that don't have the capacity to get through the times when prices dip. In fact, while doubtless political suicide, we would suggest that a "price floor" below which energy prices were not allowed to fall would be one of the most constructive measures to mitigate the coming shortage: it would promote investments in exploration, EOR, and in alternative energy technologies, as well as conservation. Of course, this is not a serious proposal, since raising energy prices would certainly galvanize opposition more readily than "death taxes" or health care. We will have to wait for prices to rise and stay high, which will most likely follow a global shortage, which will be too late to smooth the transition.

8.2.3 HIGHER PRICES

In spite of the political implausibility of shoring up energy prices by legislation, it is important to understand the role that prices, particularly cheap oil, play in promoting consumption and preventing the dissemination of the alternatives. Without some policy incentive, it appears unlikely that any of the desirable alternatives will be able compete effectively while oil is able to meet demand. It is also worth noting that any policy is likely to have an effect of increasing the price of energy, either directly or through taxes if certain sources are subsidized. Most forms of regulation tend to provide artificial support to some sources, which may not prove to be the best either in terms of energy or the environment. We need look no further than corn ethanol to see an example of this problem. It has been heavily touted as a "green alternative" and certainly healthy corn fields are quite literally green. The intensive agriculture, the spread of genetically modified hybrid monoculture crops, the water demands, and the competition with food for croplands are all serious externalities related to corn ethanol.

8.3 ENERGY POVERTY'S ROLE IN THE TRANSITION

As discussed in Chapter 7, poverty and environmental destruction are both tightly linked to lack of adequate, modern energy sources. The half of humanity currently mired in energy poverty cannot help but seek to improve their lives through increased energy consumption. In the Developed World, we cannot speak of the ethical imperative to help our fellow humans, without acknowledging that this includes helping with the transition toward energy sufficiency. What we referred to in Chapter 7 as the Energy Advantage portrays the dramatic correlation between energy consumption and quality of life for the world's poor. Some of the best news that this book has to offer is that, at least theoretically, relatively small increases in energy, targeted effectively, can support great leaps in human development. If this need is ignored, development cannot be expected to proceed effectively. We suggest that a common failure to recognize the critical role of energy in development is one of the factors explaining the historically poor success ratios of development dollars spent. While it is clear that energy alone is not a sufficient factor to support development, we argue that it is an essential factor.

It seems quite reasonable and logical that energy is needed to support all of the other facets of development. Indeed the United Nations has acknowledged this reality by declaring 2012 The Year of Sustainable Energy for All (UN 2012). If we in the affluent world ignore this need, we are likely to be confronted with it through increased competition for resources as developing countries seek to take and maintain control of their own critical energy resources. We are also likely to see it in the face of resentment on the part of those mired in poverty by the lack of adequate energy. This is likely to be especially true in places where those condemned to poverty must watch their valuable resources being taken out of their country to fuel what they will inevitably see as opulent lifestyles in the affluent world. This resentment will be exacerbated by the reality that the majority of people seldom see any of the monetary wealth generated by the export of their resources—or if they see the wealth they are only seeing it skimmed off and accumulated in pockets of their military/political leaders in many cases.

Ultimately, gross inequity is itself unsustainable. Throughout history, it has been a factor leading to war and revolution. Perhaps international terrorism is the latest incarnation of these ugly phenomena. Thus, we believe that whether one approaches the issue from the perspective of deontological ethics or of utilitarianism, sustainability dictates a transition toward a world in which most of the people have reasonable access to meeting their critical, baselevel needs. We certainly do not suggest that everyone needs a 60-inch flat screen television. We do suggest, however, that everyone needs access to adequate food, water, shelter, and basic energy. We also call attention to the Brundtland commission definition, not of sustainability alone, but of sustainable development.

To the extent that energy supports both industry and social programming, enhanced access to modern energy may catalyze development in nations if advanced in conjunction with and in support of other core activities. It can, thus, be the building block for sustained development. The kind of energy does matter. Fuelwood is not the basis for development: energy systems that yield electricity and/or fluid fuels can be. The energy mix choices and planning for the future, coordinated with development goals and programs, will be essential to making sustained development sustainable.

8.3.1 THE NEED FOR AN ENERGY LABOR FORCE

Human capacity is one of the critical needs in supporting energy development. A significant part of the neocolonial legacy was that many nations found their way to independence with a very small numbers of college trained professionals. Colonial powers had seen to this in part as a product of racist assumptions. The colonization process also found that restricting education of the local populace could be a means to maintain control. It was quite true that the educated minority of the native population commonly led the revolutions or independence movements. Indeed in the early years after independence, a large portion of the few well-educated people found themselves in government, rather than doing the work of development. Education was a top priority of many emerging nations and some have both sent large numbers of students abroad for advanced education, but have also built up their own colleges and universities. Still they had a long way to come to develop sufficient professional capacity.

The brain drain is one of the notorious challenges. Emerging nations often used limited resources to send some of their most talented young people to Europe and to America for top-notch educational opportunities. It was notoriously difficult to get those professionals to return or to stay in country after they had been exposed to the luxuries of Western life and salaries of Western professionals. Returning home and working there throughout one's professional life constitutes a clear sacrifice.

Barring corruption, even relatively senior professional people, including engineers, are paid a small fraction of what they would earn in Europe or America. But while being underpaid, they are also commonly overworked. There are demands on time and expertise from local universities, government, nongovernmental organizations, and businesses. One friend and colleague in Moçambique, upon completion of his bachelor's degree in engineering, was asked to stay on at the university from which he just graduated to assist with teaching. This was to be a short-term arrangement, until the school was adequately staffed for him to take leave and go abroad for a master's and doctorate. The short-term arrangement dragged on for nearly 20 years before he had the opportunity to take one year off to get his master's degree in the United States. While at home in Moçambique, he not only teaches full-time at the university, but he consults for the national oil company and advises two government ministries. He is very eager to go back once again to do his Ph.D. work, both for the satisfaction of personal enrichment and to expand the education that he can offer to his students. But he must not only coordinate time, but find the financial support to do it.

American doctoral students in engineering commonly have full tuition waivers, along with stipends that provide for adequate living. The rationale for this support is that the Ph.D. candidates work full-time in their advisor's lab, supporting his or her funded research program. A serious professional from a Developing Country might have to look far and wide to find a funded project at a Western university, where they can do research that not only supports their professor's program, but has value for the practical development needs of their nation. Without that institutional support, the candidates must bring their own. Some governments allocate significant money for select students from their country to come to prominent American or European universities to pursue graduate research. For a very poor country, this kind of support can be quite a drain on their limited resources, even when the need for professional capacity growth is abundantly clear. Saving up enough money to put oneself through such a graduate program is prohibitive on the salaries even of highly qualified professionals in many developing contexts.

Demands on time for the limited number of professionals are not the only constraints. Many professionals who take their responsibility seriously to support their nation's development, are not only being pulled in multiple directions, but seek some very special educational opportunities. They are likely to want to do the research and write their dissertations on applied problems. The quantum physics of developing carbon nanotubes may be very interesting science, but is not likely to address the immediate development needs of lower income countries.

Cutting-edge science, which has increasingly become the stock and trade of many research universities, will pave the way for extremely important innovations in the future, but most of these technically sophisticated and highly specialized innovations are likely to take decades to mature sufficiently to reach the marketplace. Even those that actually work in practice. Much of the innovation required for development is quite different. Rather than developing new technologies based on new scientific understandings, development requires innovative application of existing, well-proven technologies. Innovation from the perspective of Western universities and research centers typically involves scaling up from the laboratory bench, to the prototype, to pilot scale, and then to larger pilot scale and finally commercialization. Innovation for the Developing World may often involve "downward scalability" in which existing technologies are reduced from the gigawatt scale necessary for Western cities, to the several megawatts scale appropriate to the needs of Developing Nation communities. This is particularly true for energy.

Even in the Industrialized World, academic thinking may be facing a challenge in the 21st century to shift to a broader view of technologic systems, which operate at smaller scales. Since the dawn of the sciences we have been successful in dissecting problems into ever smaller components, with resulting narrowly focused subdisciplines. The challenge for sustainable development though is one of understanding and optimizing global systems. Perhaps this is a time when academia needs to find ways to train young scholars how to put puzzles back together again. Thus, academia at-large faces the challenge of developing new paradigms to train both Western scholars and visitors from the Developing World to address systems-level issues effectively. Part of systems-level thinking will be to consider contexts in which solutions are applied. Scales must be appropriate, as well as the level of technology for the professional and physical infrastructures of each setting. Practical solutions will be needed for practical problems.

The issue, of course, is not merely at the professional level. Energy development also requires a panoply of workers and craftsmen, many of them with specialized skills. For instance stove-building companies are sprouting up throughout the Developing World—filling an urgent need to replace polluting, inefficient open three-stone fires. The builders need to understand masonry and/or metalworking to build the stoves, but they also need a working knowledge of combustion processes and chimney design to vent the gases properly. Our experience working with some of these stove builders through the nonprofit AHEAD Energy (aheadenergy.org), shows some delightfully impressive talent for innovation from some of these enterprising individuals, but with minimal formal training, most of them struggle with some aspects of running their business. Whether it is a lack of understanding of finance and accounting, or strong language skills, or some aspect of the technical knowledge—many of them struggle. If we are to see solar PV or wind power broadly deployed, there must be a sufficient workforce to manufacture, to install, and to maintain all of these new energy systems. Geothermal resources, which may be among the overlooked gems in the future energy mix, will require not only engineers, but field workers to drill wells 10 feet (3 meters) to 20,000 feet (6,100 meters) deep into extreme temperatures and to complete them and operate them safely.

Many Developing Nations have oil and gas potential, which could be extremely valuable, if developed for the local markets. While not a highly labor-intensive industry, it is highly specialized. Local development of hydrocarbons will require crews of roughnecks, roustabouts, pushers, and mudmen to conduct those operations, as well as engineers and geologists and geophysicists. Even in the United States, as extensive development of the Marcellus Shale gas in Pennsylvania, West Virginia and Ohio gets underway, local residents often notice how many of the professional and labor force arrive on the scene with Oklahoma or Texas accents—even in the very birthplace of the modern petroleum industry. What will it be like as the Developing World tries to take more control of their own resources and apply them to local needs?

8.4 A BRIEF NOTE ON CLIMATE CHANGE'S ROLE IN THE TRANSITION

In this final chapter we are discussing the future of sustainable energy use in a general sense, highlighting potentials and pitfalls for each source. Much of this discussion is tied to the role that oil and gas plays and will continue to play especially in the first half of this century. It also is tied to the need for increased energy access, including from fossil fuels, in poor regions. But it will also be contingent on the extent to which climate-impacting emissions may affect global energy choices tied to environmental concerns. After all, policies to affect change in energy use patterns are often tied to the economy first, both in terms of mitigating global price fluctuations of globally traded energy commodities and in terms of a nation's ability to control energy supply. This could mean choosing fossil or renewable energy, and even specific kinds of these energies based on local availability. Secondarily, policies to change energy use patterns are tied to local and global environmental concerns. This may be tied to particulate emissions, acid rain-causing emissions, liquids spills, or greenhouse gas emissions.

Good examples that highlight how countries have addressed both of these above concerns include the relatively recent and massive shift from oil and coal to gas and wind power for heating and electricity from cleaner sources. Similarly, a shift from oil for transportation to hydrogen and electricity from natural gas, batteries, and renewable sources, seems to be generating more interest, if not momentum. If an agreed-upon tax or trade structure can be put into place, then it also may be possible to better internalize the external costs tied to fossil fuel-based pollutants to change usage patterns. However, the unexpected and substantial increases in oil and gas as the result of the shale plays in the United States and soon elsewhere, as well as a relatively stagnant global economy (at least from the perspective of the obviously unsustainable economic growth rates of the previous decades), has seemed to stymie climate-influenced decisions in favor of the more tried and true fossil-fuel-derived economic drivers. All of this points to the likelihood of fossil fuel dominance even in the face of dwindling supplies and myriad environmental concerns.

As explained in earlier chapters, we have not focused extensively on the energy/climate nexus, as the issue has been treated rather robustly in other texts. However, it is important to acknowledge that while discussing the transition, energy decisions go hand-in-hand with global environmental problems, in particular climate impacts.

8.5 ENERGY DREAMS

Our intention in this book is to present the reader with a balanced view of energy challenges and opportunities—and to provide the basic tools to understand the fundamentals of energy systems and the roles that they play in our lives. Just as it is pointless to be an alarmist, it is also counter-productive to bask in the glow of unrealistic dreams. The transition we face will not be swift and catastrophic, but it will be disruptive, as change so often is. We think that it is as important to understand that no single solution is likely to support the entire world in the sort of consumptive lifestyle patterned in the United States and Canada as to understand the inevitability of the transition.

8.5.1 EASY ENERGY TRANSITIONS

Ms. Kate O'Connor produced a song entitled "For the Love of Oil." It poignantly drew connections between terrorism, war and our dependence on foreign oil. The line "For the ease of oil, once I was pleased with oil ..." aptly characterizes much of the rise of Western dependence on the resource that can be produced in such abundant quantities, at costs that nominally decreased in terms of constant dollars through most of the 20th century. The closing stanza, though, wanders into a naïve dream state, "I'm asking my leaders to hook me up with alternative energy sources so I can find a new love to take me on long trips, cook me romantic dinners and keep me warm at night… . I'm not going to miss you [oil] at all." The song captures precisely what we want from our energy transition. All of the same benefits that we currently get from oil and other fossil fuel resources, with none of the externalities. The notion is just too easy.

That's the most challenging fiction to dispel. The last few decades have spawned a great deal of pervasive technological optimism. The computer revolution has been stunning. Many engineers and scientists are still working who learned to do math on slide rules and computer programs were seen as holes punched into cards, running on mainframes that filled small buildings. Authoring a book, such as this one, was a tedious process indeed on typewriters, requiring retyping whole chapters to accommodate minor changes. The netbook computer on which the beginning of this chapter was composed in the backseat of a car, driving home from a nephew's college graduation is more powerful than the mainframe computers that filled rooms. So, it has become popular to use the track record of computer technology to characterize a new paradigm for technologic advance. Now it is possible to start a new company, with minimal assets and to become a global force within decades, because of what Thomas Friedman calls "democratization of information, technology, and finance" (Friedman 2000). He quite correctly points to how information technology and computers have changed the playing field for businesses. So, does that mean that new energy technologies will be developed and deployed almost overnight, overcoming the issues of lag time in our information age society? Probably not.

Managing information-based technologies is one thing, but commodities-based technologies are another. Just as having a computer doesn't mean that you can overnight start an apple orchard or a dairy farm, it doesn't mean that one can start producing vast amounts of energy in the model that Amazon.com started selling books—and then darn near everything else. Computers have transformed marketing all kinds of things—and have transformed how we write this book or do research. The technology has transformed our work, enabling much more sophisticated analyses, much faster. It will no doubt, continue to support rapid development of ideas—but that is not the same as testing those ideas in pilot plants, commercial-scale operations and then taking on a large global market share. You won't be able to start a geothermal business and sell the energy on the internet. You may well be able to sell shares or futures, but not the energy itself. It will still need to be produced, converted to electricity, delivered to communities, households, and businesses. To be sure, the information management will enable testing some ideas and plant configurations, which may speed the transition. Nevertheless, a physical transformation of manufacturing and infrastructure will ultimately be required, with attendant lead times.

Perhaps, in tempering technologic optimism, it is worth looking to some negative examples. Cars come to mind. In 1980, it was possible to buy two models of diesel Volkswagens that achieved well over 50 miles per gallon on the highway. Thirty years later, surely we should be driving cars that do much better than that—while being safer and more comfortable. Even with new hybrid technologies, no car on the market in the United States currently can eclipse the fuel economies of those cars. Indeed, even if we look all the way back to the Model-T, its fuel economy was comparable to the average vehicle on the road in the early 21st century. Certainly the engines are more sophisticated, but the sophistication went into more power for bigger, heavier, and faster cars, not more efficient transportation. Technology actually does advance rapidly, but not necessarily in predictable—or even the most desirable ways. Our cars have more gadgets (and there's no doubt that some gadgets, like GPS, are nice) but those gadgets have not significantly improved our fuel economies—the improvement that will be necessary in furthering the transition to more sustainable energy systems.

We also need to recognize that the marketplace will not befriend alternative energy systems as long as the core fossil fuel systems are able to continue providing abundant, inexpensive supplies. The marketplace will respond to shortage, when it arrives, and only then incentivize the alternatives. Until then, solar, wind, hydrogen, and many others will be competing in an uphill battle. When the marketplace responds to petroleum shortages with prices that encourage development of the alternatives, it may be too little, too late to effect a smooth transition.

If you, the reader, are persuaded that the case we build, even on deliberately optimistic estimates of ultimate recoverable petroleum potential, that the shortage is likely to begin by the middle of the 21st century, then the likelihood of new technologies rising to displace petroleum easily is low. New technologies, emerging from scientific research have virtually no potential to "save our bacon" in the necessary timeframe.

8.5.2 SOLAR

It is true that solar energy reaches Earth at quantities that are roughly four orders of magnitude greater than total global energy consumption, but we currently derive an extraordinarily small part of our global human-controlled energy from solar. For example, total energy produced from PV panels stands at about 0.06% of total global consumption. Thus, harnessing enough to displace any one of the fossil fuels would require 100-fold increases in our use many times over. All the while, new global demand is added every year that exceeds all of the solar power currently produced. In effect, then, we need to continue doubling solar power production, just to keep pace with growth in demand. The "good news" is that energy prices will rise when oil's peak production is reached. Yes, that's the good news: it will help support more rapid growth of solar, and other viable alternatives. The bad news is that massive scaling up of the manufacture of solar cells will likely effect shortages of some constituent materials. And all that manufacturing requires a great deal of energy input. The transition will be much more difficult if commenced in response to an existing shortage, compared with one that is planned and undertaken while current energy resources remain abundant.

A few decades before useful energy production from solar power realizes clear gains beyond the energy required to make all of those new PV cells, a tremendous manufacturing effort will be required. The transition can happen, but is not likely to be smooth or easy or immediate. Consider that for solar energy to meet 20% of global energy demand in 20 years, it would require manufacturing 100 million 200-watt solar arrays per day! To maintain a 20% market share, the solar cell production would need to continue at a comparable rate, since solar arrays have realistic useful lives of something around 30 years (so some of the ones produced in the first year likely will need to be replaced after the goal would be met in those first 20 years). That is a huge manufacturing venture, which will certainly incur massive energy costs of its own—and probably some manufacturing, transportation, and disposal costs.

8.5.3 UNPROVEN TECHNOLOGIES

Nuclear fusion is an example of the energy fantasy that a new technology will ride in like a knight on a white charger to meet all our needs. The scientific research is clearly valuable in its own right, but to place our confidence for energy security on the back of unproven technology is ill-advised. It is comparable to placing our hopes to meet the next global crisis on colonizing other planets because NASA generates a lot of good science. NASA does, and humans may well travel to distant planets someday, but not any time soon. The same is true of fusion technology. If the task at hand is to vaporize a city, fusion is the go-to technology. If not (and we sincerely hope not), controlled fusion remains a holy grail of technology: ever tempting, but just as elusive. We certainly support ongoing fusion research—it generates good science and it may obviate energy worries for our grandchildren (or their grandchildren). But we cannot expect it to solve the imminent shortage.

Oil shale is also a highly problematic resource on which to bet our futures. Unlike shale oil and gas, which tap liquid and gaseous products, the oil shale resource is based on extracting from rock solid organic matter that has not yet matured to produce gases and liquids. There is excellent evidence that a tremendous amount of oil shale exists in the world—no doubt considerably more than petroleum. But acquiring it presents tremendous challenges as well. The only technology that has been proven to be successful at producing commercial quantities of syncrude from kerogen has been mining, crushing, and retorting. This process has low net energy efficiencies (positive, but low) and significant environmental externalities. At the scale of 10,000 barrels of oil per day (comparable to a single very good, Middle Eastern oil well or a little over 0.1% of global petroleum demand currently), using a very high grade oil shale formation, with federal price guarantees that were equivalent to more than $100 per barrel in 2012 prices, the producing company still lost a billion dollars. *In situ* processing may be more cost-effective and would certainly have less environmental costs. But, that is a new technology, unproven at the time of writing this book. It may develop rather quickly, but is not a guaranteed success. And, since it depends on burning product in ground to gasify or liquefy more product, its net efficiencies are limited, only a portion can be recovered. Of course, the same is true of petroleum, but the oil left behind is still there and more may be extracted with EOR. The kerogen burned is lost forever. Oil shale might become a major contributor, based on *in situ* processes, but it will take time. The real potential of this resource may be in tapping the partially matured shales from which oil and gas can be produced directly.

8.5.4 RIDICULOUS TECHNOLOGIES

As with any dreams, some have the semblance of truth, while others are the wildest of fantasies. These wild dreams include all manner of perpetual motion notions: those that defy the laws of thermodynamics. While any energy source must yield more energy than humans put into the process, the resulting energy must always come from somewhere: the fusion reactions going on in the sun; the uneven heating of Earth's surface from solar radiation; falling water; the heat of Earth; or the chemical bonds of carbon-based fuels. If one cannot identify the source of energy, it is almost certainly a pipe dream. Any technologies that speak of deriving energy from water must offer some explanation to what reaction takes place to yield useful energy. No energy transformation can yield an increase in energy. It is not, for instance, possible to pump water up one pipe and obtain more energy from water falling down another pipe than was used to pump it up.

Another class of wild dream is that some hidden technology is just waiting to revolutionize our use of energy. In the energy crises of the 1970s, urban legends proliferated about a carburetor that would allow any car to get 100 miles per gallon. This was a clear physical impossibility: the carburetor only mixed fuel, so a perfect carburetor could only improve the combustion efficiency, which was relatively high to begin with. Most of the inefficiencies of automobiles relate to air resistance and friction losses. There was not nearly enough room for improvement to quintuple fuel efficiency (Kraushaar and Ristinen 1993). However, many people believed it—all they needed was to hear the conspiracy theory that some unnamed oil company had purchased the patent, in order to keep the great invention secret. That, by the way, is not how patents work. The patent process requires making the principles of the invention public. It merely protects the discovery for the inventor to develop, free of competition for a finite period. So, when you hear conspiracy theories of this sort, they are almost certainly false. Indeed, those theories do not even hold out hope for any viable option for the path forward; they merely seek enemies to blame.

8.6 COMPARING THE OPTIONS

The choices will be neither simple nor straightforward. No single alternative energy source is likely to be able to replace a very large share of the energy currently provided by petroleum. Every energy source has constraints based on physical limits, technologic development, environmental impact, economics, and social considerations. Each of these factors varies depending on context. Contextual issues to be addressed include resource availability, geopolitical issues, infrastructure, time horizons, and consumption patterns. Solutions for New York City or Boston will be very different from those for upstate New York, Germany, or Spain, let alone for Moçambique or Uganda.

So how do we craft a plan for smooth transitions to increasingly sustainable energy systems? Viable planning rubrics must not merely optimize the value added by resource mixes, but account for changing circumstances with time—and the solutions must be practical. It does not suffice to say that nuclear power has the technical capacity to take over a large part of petroleum's market share for many decades. We must consider all of the practical ramifications, including the social and political acceptability of disseminating and operating nuclear power facilities throughout the world. Nor does it suffice to say that all of the electricity demands of the world could be met by solar PV installations covering less than 144,000 square miles of arid land with solar cells (EcoWorld 2006). That's a little more than the size of the state of Arizona or about a fifth the size of Algeria. We must also consider how that power is distributed throughout the United States and stored as well as the economic inputs required and the energy inputs required. We have already discussed the enormity of the manufacturing task required to achieve a mere 20% of global energy demand. How much energy will be spent to make all of those PV panels and where does it come from? Furthermore, even if experiments to produce a sustained net energy gain from controlled fusion are stunningly successful by the time this book goes to press, it will probably require two to four decades before fusion could take a significant market share.

Time is an important factor. For the stock-limited, depletable energy sources, physical limits on the availability of the source leads to a sustainability problem. On the other hand, sources or technologies that are not fully matured are limited in terms of the lead time it will take in order for them to be able to assume a significant market share. We need to balance our national and global energy portfolios to ensure that sufficient energy is available throughout the foreseeable future. Some resources, such as gas from shales, have peculiar production timelines: a newly drilled well, with a massive hydraulic fracture stimulation, will produce prolifically for a short period of time and then begin a very steep decline, but level out at a much lower production rate and continue to produce at that lower rate for many decades. In planning energy transitions, it will be vital to understand how each viable resource will produce across the time dimension. Much of the challenge of a transition toward more sustainable energy systems will be how societies balance the decline of existing dominant energy supplies with growing demand and the advent of each alternative energy system.

8.7 NEW LIFESTYLES AROUND SUSTAINABLE ENERGY

In light of the magnitude of energy transitions required and the enormity of the gap between the declining production of oil and the aggregate production of the nondepletable alternatives, it will be imperative to find ways to conserve, especially in highly consumptive nations. We believe that the nature of what we call the Energy Advantage in Chapter 7 demonstrates that there is a good deal of energy consumption in much of the Developed World that does not add to quality of life. Changing toward less consumptive lifestyles may actually enhance qualities of life—the same, of course, is true of consuming less food for many of us.

Some of that transition to less consumptive lifestyles will be a matter of opting out of some discretionary uses, such as: driving short distances; idling cars in parking lots; leaving appliances on when not in use; relying heavily on electronics for entertainment. Some of the transition may require some sacrifice: driving smaller, more efficient vehicles, selecting appliances for efficiency rather than luxury. Some will no doubt require societal changes. Suburban and exurban sprawl will likely be seen in future years to be wholly unsustainable. Not only will long commutes in individual cars become prohibitively costly for many as fuel prices increase, but sprawl makes viable public transportation exceedingly difficult. This transition may be painful, as it is unclear what is to become of the expensive homes scattered in deliberately remote exurban communities. Unlike when urban flight led the upper and middle classes to the suburbs, leaving the poor to occupy inner city tenements (some of which were formerly upscale residences), there will be no opportunity to convert the remote homes into tenements. Who would live there?

Indeed, the transportation sector in the United States is a prime target for the transition. It alone constitutes 1/8th of total global petroleum consumption. In spite of the large opportunity for conservation, it won't be easy. Much of the consumption is a function of our living patterns. Since exurban living cannot simply collapse into urban centers, it will likely take decades to reform American living toward more centralized communities. New car purchases will probably shift dramatically toward more efficient vehicles, but it will take time for auto makers to shift their production lines as dramatically. The car makers will need to be confident that demand is fundamentally shifting toward energy efficiency. Even as people buy more efficient new cars, the less efficient ones do not instantly disappear from use.

Conservation is not a sufficient solution to the energy challenge, but it is an essential component. The flow-limited, nondepletable sources like solar and solar-derived energy do have practical limits to their production rates. These limits include the constraints on the use of other resources: especially land and water. Without conservation, the target energy production to meet the world's needs would be too high for the nondepletable sources to meet without some unforeseen breakthroughs. Also, an inevitable piece to the conservation puzzle will be achieving stable global population levels. Growth has been and will continue to be an intractable component in forcing demand higher, until populations stabilize. Of course, most of the population growth in the early 21st century is in the Developing World. This should not be taken as reason to discount the importance of population stability in affluent nations, though, because every person consumes so much more energy (as well as other resources), that every additional person living in affluence is the equivalent burden on planetary resources of 50–100 people in the Developing World.

The issue of restraining population growth is a highly contentious one, with a flavor of governmental policies impinging on reproductive rights. The one-child policy in the People's Republic of China was viewed in much of the Western world as draconian and seems to have helped to mobilize resistance to any form of governmental policies to slow population growth. While the difficulties of legislating family size in most cultures are probably overwhelming, we would suggest that the population piece of the puzzle cannot be ignored. As noted earlier, some of the extreme points advanced by those opposing any notion of limits to growth are patently absurd. The universe itself sets inviolable limits, let alone our galaxy, our solar system, or Earth to which our existence is constrained for the foreseeable future. It would take only a little over 1,100 years at 2% annual population growth rate for the entire mass of the planet to be taken up in the bodies of living humans—utterly impossible. As seen in Chapter 5, even at a more modest 1.5% annual growth rate in human population, it would only be a couple of centuries before humanity would occupy the habitable portions of Earth's surface at an uncomfortable density, with an average of 324 square feet (30 square meters) of living space. We cannot overemphasize the ultimate reality of limits. If population continues to grow at any exponential rate, it makes all other efforts toward enhanced sustainability futile.

In general, we must find ways to limit demand—especially growth in demand—to reflect the limits to growth set both by the physical universe and by reasonable expectations for the rate of development and deployment of alternative energy resources. We will pay a greatly increased price for energy once the global petroleum peak is reached and that will be a heavy burden for wasteful consumption, even with new energy resources gradually taking larger market shares.

8.8 OPTIMIZED ENERGY MIXES FOR SPACE AND TIME

The global transition will not be a single nor homogeneous step. Solutions will need to be optimized for each context—and sufficiently agile to adapt to changing realities. There are factors that affect energy mix optimization: resource availability; socio-political issues; physical infrastructure; consumer needs and preferences; cost; and environmental impact. Arid and some low latitude regions have the most consistent solar power. Some have virtually no wind, which is often abundant in coastal regions or mountain ridges. There is certainly uneven distribution of the fossil fuels—though perhaps less so than is commonly believed. Sugar cane is a much more productive ethanol feedstock than corn, but requires warm climates. Conservation opportunities are the largest in the most consumptive countries. In the less consumptive countries, efficiencies can help to achieve benefits with less additional energy use, but it seems unavoidable that they must increase their energy consumption.

To the extent that natural gas is locally available, it is very difficult to outcompete: especially for heating needs, including cooking. It burns cleanly, efficiently and is extremely cheap. It can also be very economically attractive for electricity generation. Gasoline or diesel offer such impressive energy densities that they will be hard to replace: compressed natural gas seems to offer one of the most likely near-term alternatives, which would also offer significant reductions in carbon dioxide and classic air pollution emissions. (Hydrogen may be able to pick up market share in this regard, as fuel cells can utilize natural gas with similar or better efficiencies than typical compressed natural gas engines.) As long as they can be produced in abundance, their values will be hard to replace. But that value will begin to decrease with time, as they do deplete. This factor also raises questions about the highest-level use of each fuel, in light of its finitude. Natural gas, for instance is very abundant in the marketplace—and likely to become more so—at the time of this writing. Expanding its use in vehicular transportation and stationary power generation would both offer some benefits for air pollution and greenhouse gas emissions, quite economically. But, will the near-term benefits outweigh the opportunity cost of having less available once depletion does become an issue? How do we compare the near-term and long-term benefits? How should we value the cost of committing to more, or new, infrastructural support for resources that will begin to decline in the foreseeable future? What amortization period makes sense for such infrastructural investments? We believe that a dynamic model must be developed that plausibly compares the disparate costs and benefits as factors change over time.

8.8.1 USING EVERYTHING, AS WE ALWAYS HAVE

What energy resources should we develop for the future? That is one of the most significant questions to ask. The simplest answer is everything that works. A corollary to that is not to waste time and money on things that won't work. We should develop every energy source technology that can realistically contribute to the global energy mix because no single one is likely to be a sufficient solution. The world is likely to need a much broader portfolio of energy mixes than it has ever had before. If the coming transition were not so urgent, it might not be important to avoid dead ends. This is one of the ways in which the "moon race" analogy to the energy transition fails. The goal of landing on the moon was scientific and political. The security of human societies did not depend on reaching that goal. Unfortunately, the energy transition needs to proceed as quickly and effectively as possible in order to minimize socioeconomic disruptions. Inevitably some sources and/or technologies that seem to have real potential will still fall by the wayside in the path to development and the marketplace. There are, however, some avenues that can be foreseen not to be worth pursuing and should be avoided. The energy mix of the middle 21st century is likely to include solar space and water heating, solar PV, concentrated solar power, large and small-scale wind, geothermal, increased hydropower (perhaps especially small-scale), and waste-to-fuel processes. It should actually go without saying, but public and even scientific discourse too often overlook the reality that the fossil fuels will continue to provide a majority of global commercial energy and dominate the energy mix for decades to come. It may well take a century for the noncombustion alternatives to take over a majority of the global energy markets. Even then, bear in mind that the aggregate fossil fuel contribution is likely to be larger than any one or two alternative energy sources. Oil, gas, and coal are not likely to go away anytime soon. Indeed, the collective solar technologies that have been developed and deployed over the last 30 years have just barely taken a 1% market share of global energy consumption.

What about nuclear? Fission reactors already account for the majority of noncombustion energy production in the world. The technology is proven, the resources are relatively abundant, and breeder reactor technologies seem technologically poised to expand the viable nuclear feedstock to make the resource truly superabundant. Nuclear fission does have the technical potential to expand to meet the gap between supply and demand as we pass the inflection point on the growth side of the global petroleum peak curve. And it produces no carbon emissions.

Its development is largely constrained by fears. Within the affluent world, fears of nuclear accidents and of the hazards of radioactive waste constrain broader deployment of nuclear fission. This is especially true in the United States, where no new nuclear power plants were constructed for half a century. Just when the tide seemed to be shifting and the Obama Administration began to shift public awareness in the United States to consider that nuclear fission might play an essential role in our energy portfolio, the 2011 earthquake in Japan, and the subsequent Fukushima nuclear disaster may well have set back nuclear energy development in the affluent world by decades. It does demonstrate that even with good engineering and safety protocols, it is essentially impossible to prevent all possible hazards. Large-scale operations have the potential for large-scale catastrophic failure. The core fear for nuclear is that nuclear power has the potential for a failure to impact civilians for miles or hundreds of miles around the plant. While arguments can be mounted to dispel most of these fears, fear itself is not rational—but it is real and a powerful motivator.

Nuclear waste disposal remains a limiting factor in the expansion of nuclear power. It is true that high level radioactive waste is dangerous for centuries—of course, chemical toxins resulting from other processes may last forever. Somehow it seems that nuclear radiation is scarier than cyanide or arsenic or other chemical toxins. Since radiation can penetrate solid containers, and human flesh, it is important to determine how best to dispose of it.

The other serious constraint to nuclear power relates to the proliferation of nuclear materials and technology. There is a certain commonality between nuclear power and nuclear weapons. The real technical challenge in making nuclear weapons is in sufficiently processing and refining the fissile materials. Nuclear fuel is not weapons grade, but it may be reasonable to fear that the processing to achieve fuel grade may be a step toward achieving weapons grade processing. There is also concern about nations using a nuclear power industry behind which to hide nuclear weapons development. For this reason, it seems exceedingly improbable that nuclear power will be an effective solution for the half of humanity in the Developing World.

Nuclear fission offers significant potential, but we doubt that is likely to respond to a crisis with rapid growth, while at the same time, socio-political concerns are likely to prevent its broad deployment in many Developing Nations before the middle of the 21st century. (We note that with over 20 reactors currently under construction, China does appear to represent the exception to this likelihood.) It may become a larger global player in the longer term, as technology for safe operations and waste handling increase and public perceptions shift.

8.8.2 CONTEXT-BASED SOLUTIONS

Much of the work of a transition to more sustainable energy use in society will not be high-tech. It will rather involve thoughtful efforts to find new applications and to improve existing, even mundane technologies. It may be tempting to dream of technology riding to the rescue of our casually consumptive lifestyles in the affluent world, but that is not likely, especially in the time frame required. When one remembers the already present energy crisis of the Developing World, the observation becomes even more pointed. They need access to a range of existing technologies to support prudent development of their own energy resources to meet their own needs. Technologic development often entails "scaling up" from the lab bench, to a prototype, to a pilot facility, then to full, commercial scale. In the Developing World, though, we again suggest that "downward scalability" may be the more valuable approach. For people consuming 1%–2% as much energy as Americans, a challenge is to scale technologies to be appropriate to local needs and to be locally operated and maintained. Local control of right-scaled projects is also likely to be an important antidote to corruption.

8.8.3 LOCAL, DECENTRALIZED ENERGY DEVELOPMENT

Just as the need for a transition is compelled, at least in part, by the ultimate limits of resources, their availability is an essential aspect of sustainable energy development. The extent to which a nation, region, or even locality builds its development on locally available resources both mitigates efficiency losses in moving the energy to the consumer and reduces risks of supply disruption. The case for local, decentralized energy production and consumption appears strong. We think that will be especially true in the Developing World, where the cost of infrastructure to support broad distribution from large, centralized energy production adds to the enormous capital costs of large-scale projects, representing a major obstacle.

Some researchers are working on means to make the long-distance transmission of electricity safe, efficient, and cheap. Superconductivity is the primary avenue of inquiry. At extremely low temperatures, some materials lose most of their resistance to conducting electrical current, making them "superconductors." Some "high temperature" superconducting phenomena have been noted and there is an application being tested on a stretch of power grid less than a half mile long, by the Long Island Power Authority (GENI 2008). It is still worth noting, that even the application that is referred to as "high temperature" involves refrigeration. In practical terms, we believe that long-distance transport of vast quantities of petroleum is and has been problematic. Long-distance transmission of enormous quantities of electricity may prove to be impractical. This point may change with more research in coming decades, but for the time period through the middle of the current century, it does not seem likely to change.

Furthermore, extremely large-scale operations are also extremely capital intensive. The economies of scale are dependent on having sufficient capital to make the investments in the mega-scale facilities. These investments are likely to remain prohibitive for the most economically disadvantaged nations. Small-scale, localized energy projects appear to be the only ones likely to support development in many communities in the Developing World. In the Developed World, localized energy production must compete against energy systems with established infrastructure. Nevertheless, some do advocate for decentralized energy for affluent nations as well. This can be helpful to the alternative energies, which are often produced at relatively small scales and have some potential to be adopted for home, rooftop applications. This kind of application eliminates the land requirements associated with large, centralized power plants.

8.8.4 CONSERVATION

We believe that conservation efforts will be an essential part of the transition. While we acknowledge the point from the so-called "rebound effect" that conservation and efficiency improvements cannot be expected to reduce global energy demand, we believe that the transitions rely on those improvements, as past transitions of industrialization and economic development have also. The need for energy in the Developing World will—and should—drive growth in global demand. The transition to a more socially sustainable world requires that 3.5 billion gain access to adequate energy. Without conservation on the part of the world's affluent, increased access to energy for the other half of the world would be virtually impossible.

But conservation is not just about improving efficiency. It is not merely buying the newest "Energy Star" refrigerator, or the newest hybrid car. It is about making choices in how we consume—and why. Why are we consuming energy? What are we seeking to get from it? Mobility? Comfort? Entertainment? Does a large SUV transport us any better than a modest vehicle? We need to become deliberate in deciding to consume energy for a purpose. When we do, we may well find that we are living just as well—perhaps better—with less. Less can be better.

Indeed, energy so undergirds all we do that we must consider all of our consumption choices. When we choose to eat more locally grown foods, from lower on the food chain, what backlash of increased consumption can there be? When we choose to live closer to our work, to use public transportation, to buy homes of a size that matches our families, we will consume less energy. When we make deliberate choices about what we consume and how we consume it, we are likely choosing more sustainable lifestyles.

But, as we asked in Chapter 1, "who will do it?" Certainly, not everyone will choose more conservative lifestyles. Indeed, none of us are likely to make all of the most conservative choices. Some of us like movies and will find there is added utility in a large flat screen TV. Some truly enjoy sportscars and will gain pleasure from owning and driving one. The point is not to say what you should do. The point of this section is to suggest that deliberate choices may well reveal that some of our consumption does not add significantly to our quality of life. Just as being more deliberate in eating may tell us that the third trip to the buffet does not really add to our pleasure. More efficient and thoughtful energy consumption is the goal. Nevertheless, the world will need a great deal of energy for a long time to come.

8.8.5 EVOLVING ENERGY MIXES

If we proceed with a successful transition toward sustainability, it seems clear that it cannot be a single step, nor met by a single resources will be required to meet the world's needs. Although stock-limited, fossil fuels must inevitably continue to provide a great deal of the world's energy mix for decades to come. The viable alternatives of wind, solar, geothermal, and small-scale hydropower will need to expand even more rapidly than they currently have been, just to offset the shortage resulting from peak petroleum production. At the same time, the world needs to see more rapid growth in global production of modern energy to provide for the needs of the rest of the world.

The transition will be a dynamic journey, rather than a singular step. We will need to optimize energy mixes by context, but plan for those mixes to change as some resources deplete and alternatives grow with time. It will require continual reevaluation, but some planning is possible a priori. We can know with some certainty that the role of petroleum will have to diminish by the middle of this century. In spite of common perceptions, we have good reason to believe that the petroleum decline will be gradual, but even a very small percentage decline will lead to large shortages because of the very large production rates.

Fusion, while having huge potential for the long term, cannot realistically be assigned any positive value before the middle of the century. Even then, we would suggest that its potential should be heavily discounted for the uncertainty associated with its development.

Biofuels may play a role, especially from waste streams. We believe that these are valuable in that they do not compete for land and other resources—rather it is a win-win venture in putting waste products to good use. This may include some cellulosic ethanol sources. Algae crops have some potential to grow biofuel feedstock. It appears to have an extremely high yield, with minimal land use and the potential to be coupled with carbon capture efforts. Nevertheless, the potential contribution must be significantly discounted until the technology has been thoroughly demonstrated at practical scales. Sugarcane can be successful where it can be grown, although transforming millions of acres to new sugarcane production is likely to have serious environmental consequences. We would not expect it take any significant global market share.

Other biofuels, generally seem to hold very little promise, in our estimation. The net energy produced by ethanol from corn, which has received so much attention—and so many subsidies from the United States government—appears to be too small to be considered a serious energy resource. The competition with food and the environment for arable land appears to be a serious issue, as well as the environmental challenges associated with massive scale monoculture agriculture all appear to bring serious externalities. Firewood and charcoal use will no doubt continue to provide a large share of the energy for the Developing World, but we would suggest that the transition away from raw biomass dependence should be one of the highest priorities for sustainability: to support the environment, reduce fatalities, and support sustainable development. Modern biofuels will continue to add marginal energy production, but may actually lose some market share in the medium term as the real value limits gain broader recognition—this may be offset by enhanced utilization of waste streams. The overall share of biomass in the global energy mix should decrease, at least until some exotic technologies, such as algae production take larger market shares.

The growth of solar and wind power, will continue at the highly impressive rates that they have both established in recent decades in order to take up significant market shares. If each were to reach 10% of the global market (for a combined total of 20%) it should be viewed as a major success for the development of these energy systems. It is important to recall that solar energy is not limited to the popular images of PV cells.

Geothermal is one of the few sources that seems to have more potential than is commonly discussed. The heat of Earth's interior is enormous and increasing due to the decay of naturally occurring radioactive isotopes. Hot dry rock investigations have been largely shelved since the 1980s, but we suggest that there may be renewed interest in light of the combined higher prices that will attend the global peak in petroleum production and the advances being made in induced fracturing of formations and horizontal drilling. It appears from our analyses that the new technologies required are in the category of engineering development, as opposed to needing major scientific breakthroughs. Since the work has been largely on hiatus for a quarter of a century, it may take significant time. Since the new technologies are not high risk, but are not yet developed, we would suggest that it may be possible for geothermal energy to expand dramatically, still seeing 5% of global energy production as an optimistic target by the middle of the century. We think that it has much higher potential for the longer term, in light of the immense resource base, the relatively modest and controllable externalities, and the demonstrated basis of current production.

Hydropower is likely to maintain its current market share and perhaps gain. The environmental and social disruptions attendant with new "big dam" projects seems likely to curtail expansion of that sector. However, the potential for much smaller scale hydropower projects suggests that there may be some expansion of hydropower globally. Small-scale, run-of-the-river types of projects seem to have considerable potential to meet the needs of people in Developing Nations. Without large new additions, though, the already substantial hydropower consumption seems likely only to maintain its market share, perhaps expanding it slightly.

Systems are likely to evolve toward more decentralized or distributed energy. There is considerable impetus behind the transition toward less centralized energy systems. As we've discussed, this seems to be nearly an imperative in the Developing World. It can spread out the investments, rather than incurring the huge front-end capital costs to build large centralized projects and to extend the grid around the countryside. It can serve as a "nucleation site" model, which we argue is a better match for how power developed and spread in the Industrialized World. For the affluent world, the transition toward decentralized energy will face greater challenges, since it competes, in effect, with the established infrastructure and human capital supporting large-scale, centralized energy systems. These have the advantages of economies of scale and the flexibility of being able to shift energy from one market to another dynamically. Of course, this latter advantage is one of the challenges of centralized systems, too, as we have seen in the cascading collapse of interstate electric power grids around the turn of the millennium. Small-scale, localized energy and production systems can have fewer transmission/delivery losses, which may help to offset the economies of scale—and their local nature makes it easier to view energy systems in their entirety. We think that small-scale, localized energy production and consumption will be the most successful means to extend the Energy Advantage to the three billion people lacking access to modern energy. The shift toward less centralized energy systems in the affluent world seems likely to proceed more slowly.

8.9 BRIEF SUMMARY OF AGENCY AND INDUSTRY FORECASTS

To add some "teeth" to the generalizations of this final chapter, we end by reviewing how the major agencies and oil companies currently view future energy use, at least through the first half of this century. As MacKenzie (2010) summarizes in an article for the Financial Times, the most recent forecasts coming out of the Energy Information Administration (EIA) and the International Energy Agency (IEA) focus on one of two concerns: the future price of oil, or controlling carbon dioxide equivalent (CO_2e) emissions. EIA focuses on the price of oil as the driver of future energy use, while IEA focuses on CO_2e, tied, in part to the interests of their constituents in the United States and Europe, respectively. In reality, both of these foci are related. The higher the price of oil, the more likely it is that less of it will be used, thus reducing CO_2e emissions. As such, both EIA and IEA produce multiple energy use scenarios based on different levels of accepted atmospheric CO_2e concentrations or different prices of oil. We will focus on IEA, as it has released the 2012 edition of the *World Energy Outlook* recently and no new report from EIA will come until well into 2013.

In its reference, or "Current Policies" scenario, forecasts assume little will be done to reduce CO_2e emissions and that the price of oil will stabilize at slightly higher than current price levels. In this scenario, we can expect much of the same, with growth rates of oil and gas remaining at about 1% to 2% per year, while renewable energy growing at much faster rates, though overall contributions will remain small in comparison to the fossil fuel strangle hold on energy consumption. In this scenario, global oil demand would exceed 99 million barrels per day in the year 2035, and unconventional and deep water oil will likely help boost the non-OPEC supply in the short term, but the world will rely on OPEC heavily after 2020, much from Iraq of all places (IEA 2012). Also, and perhaps more frightening from an across-the-board pollution perspective, IEA predicts that global coal demand will rise 2.6% annually over the rest of this decade, and may catch up to oil as the most consumed energy source (Bloomberg 2012). In more aggressive policy scenarios, energy efficiency, through conservation and technology changes, can greatly reduce increasing energy demand in China, India, the United States, and Europe, however a concerted effort would be required to have this happen.

As far as the oil companies are concerned, Shell, BP, and Exxon Mobil all seem to foresee a "business as usual" energy future (MacKenzie 2010). Though this attitude would appear to be at first blush self-serving, it is probably not unreasonable. Infrastructure is very slow to change, thus paving the way for continued use of oil, gas, and coal. Even in developing regions, where infrastructure is scant, it is likely that oil, gas, and coal infrastructures will be incorporated into the energy mix. Moreover, a globalized economy requires ever more transportation energy, derived from petroleum. In Royal Dutch Shell's publication, *Signals and Signposts*, world energy demand will grow from about 500 Quads in 2010 to well over 700 Quads in 2030, with developing countries representing 70% of this demand (Shell 2011). And the mix will still have oil, gas, and coal represent the majority of the energy supply at about 70%–80%. That means a lot of oil and gas will need to come online in the coming decades. It also means that renewable energy is starting to grab a more meaningful share of the energy mix, however not fast enough to dominate until well past 2050—unless something drastic happens soon.

One last set of points that should be discussed involves the IEA observation that the recent growth of oil and gas in the United States seems to drastically change energy flows in the coming decades. Granted, it is hard not to think this is a relatively short-term phenomenon, however, based on new United States production estimates, the IEA predicts that the United States will become a net exporter of gas by 2020 and is net energy self-sufficient by 2035 (IEA 2012). This is wildly different than what had been predicted even three to four years ago. Moreover, coal use in the United States will become greatly reduced thanks to all of this gas being produced in the shale formations. (As noted above, elsewhere, coal use will continue to rise, especially in China and India.) Also, North America as a whole becomes a net oil exporter, due to increasing unconventional oil production, including from the tar sands, while oil production from Saudi Arabia and Russia will decline at least in the short term. Michael T. Klare correctly points out that if Saudi and Russian oil production is on the decline, this forecasted oil shift appears to coincide with a shrinking world oil supply (Klare 2012). However, it seems likely that the hydraulic fracturing/horizontal drilling techniques that have been developed in the United States will also work in Saudi Arabia and Russia, which makes long-term predictions of a shrinking world oil supply, based on recent events, tenuous.

After all, we believe that more oil and gas will be discovered in the era of peak production rates, even if it cannot keep pace with historical growth in demand. And while current growth rates will become untenable by mid-century, much oil will still be available. Finally, the IEA predicts that renewables could also continue its push as a major source for power generation and that efficiency could displace 20% of global energy demand by 2035; however, this is contingent on subsidies and strong political will. In the face of the political pressure to maintain 20th century economic growth patterns, this may be difficult to achieve.

8.10 SO, WHAT IS THE PATH FORWARD?

What can we reasonably expect and how, then, can we plan for a more sustainable energy future? First, there can be no argument that we are depleting the stock-based energy sources. When to expect shortages is arguable, but, using an estimate of ultimate reserves for petroleum three times the size that most peak oil theorists use, we still suggest that there is good reason to believe that the peak will come before mid-century. Our own analysis yields 2034 as a hypothetical time for that peak, but we certainly do not attach any confidence to that number—only to the reality that it is before mid-century. That suggests that other energy sources will need to be taking up some of petroleum's market share within 20–40 years of this writing. We suggest that is a serious challenge.

The most popular alternatives—solar and wind—have been increasing their global production rates dramatically and will probably continue to do so, but that increase is barely enabling them to hold their meager combined market share. Even extrapolating their record-setting growth rates does not have solar and wind providing as much energy by mid-century as a projected difference between declining petroleum production and extrapolated growing global demand. Their growth is important, but does not appear sufficient to meet the looming crisis alone.

Coal can do it, but with noticeable additional environmental impact in the near term. "Clean coal" *in situ* processing techniques might permit coal to expand its market share without exacerbating environmental problems. We offer the caveat, though, that expanding coal operations thus far have moved into greater environmental impacts more than reduced impacts: such as the mountain-top removal form of strip-mining. Still, there is some hope for a resource that can respond rather quickly.

Natural gas can expand in the near term, though its global peak is likely to follow oil's within a half century, especially if its use expands dramatically. The new developments in producing gas (and oil too) from shales previously deemed nonproductive holds the greatest promise for pushing the crisis further out in time. These resources should almost certainly be pursued. Like any other development, the potential for externalities should be carefully monitored and those that are found to be substantial should be assiduously addressed.

Nuclear fission reactors are proven technologies, but fraught with problems—both real and imagined. Nevertheless, as we seek to understand energy systems, public perception cannot be ignored. It can grow and probably will—after a crisis forces energy prices up. It will probably help to serve our medium to long-term needs, but does not appear likely to rise to meet the imminent shortages.

So, there will be no single solution. To all of these energy sources and others, conservation must be added. Nothing will be able to meet exponential growth demands indefinitely. We need to understand what we want from energy and seek an optimal balance between the combined internal costs of energy and the benefits we derive from it. Understanding how the benefits saturate at some level should enable us to be more deliberate in our choices of how much and how to use energy.

We believe that optimizing evolving energy mixes over time, based on locally available resources, infrastructure, human capacity, and needs will require systems-level understanding. The problem of planning a transition toward more sustainable energy systems will need teamwork: teams more like football teams than track teams. They are both great sports, but on a football team, one needs to play their position well, but also needs to have a general understanding of what the other players on his team and his opponent's team are doing. All of them need to understand where the ball is on the field and know something about the competition. A track team, on the other hand, is much more like prevalent academic pursuit. As former shot-putters, it would have been possible to get off the team bus, without knowing anything of the competition, walk to the ring, warm up, without paying any attention to what anyone else on the team was doing, get in the ring and heave the shot four times. Knowledge of what the rest of the team was doing had no effect on an individual's performance, nor did knowledge of one's opponents. (It was more fun to be a part of the team and watch other events before and after one's own and even to know what to expect from the other team—but none of that truly affected one's performance.) Much of academic disciplinarity tends to be like a track competition: rather isolated. In conceiving of the most effective transitions, the world needs people who do understand the context, and how different resources work together and compete. That, we believe, requires some new kinds of training.

There is good news. First, we believe that there is good evidence to expect the petroleum peak to be something of a plateau, followed by a long, gradual decline. It will be nothing like falling off a precipice. Furthermore, taken collectively, it appears possible to assemble energy mixes to meet the combined challenge of increasing access to modern energy for half of humanity, in the face of declining global production of the current mainstay energy source. It will, though require a mix of energy solutions, which will change over time, as some deplete. We cannot expect to divorce ourselves from the depletable energy sources in the near or medium terms.

Finally, we must understand the whole challenge, we believe by taking a global systems perspective. We must understand the need for increased access to energy for the Developing World to progress on a development path. At the same time, we must understand the limits to the benefits of increased energy consumption. We must be able to optimize energy mixes by context over time. It will be a great challenge, but it can be done, if undertaken thoughtfully and with dedication. Good luck to us all.

REFERENCES

AHEAD Energy Corporation. 2012. http://www.aheadenergy.org/index.html

Decanio, S. 2006. "Market failure." In: The Encyclopedia of Earth. Washington: Environmental Information Coalition, National Council for Science and the Environment. http://www.eoearth.org/article/Market_failure

Ebenhack, B. W. 2010. "Preserving and Extending the Energy Advantage." Society of Petroleum Engineers 2009–2010 Distinguished Lecturer Program. http://www.spe.org/dl/2009.php

EcoWorld. 2006. "Biofuels vs. Photovoltaics vs. Biomass." http://www.ecoworld.com/energy-fuels/solar-power-biofuel-vs-photovoltaics.html

EPA (U.S. Environmental Protection Agency). 2012. *Causes of Climate Change*. http://www.epa.gov/climatechange/science/causes.html

EIA (Energy Information Administration). 2010. Energy Consumption by Energy Source, 2001–2005. http://www.eia.doe.gov/cneaf/solar.renewables/page/trends/table1.html

Friedman, T. 2000. *The Lexus and the Olive Tree: Understanding Globalization.* New York: Anchor Books.

GENI (Global Energy Network Institute). 2008. "Superconductor Cable gets Energized on Long Island." http://www.geni.org/globalenergy/library/technical-articles/transmission/cleantech-group/superconductor-cable-gets-energized-on-long-island/index.shtml

Humphreys, M., J. D. Sachs, and J. E. Stiglitz, eds. 2007. *Escaping the Resource Curse*. New York: Columbia University Press.

IEA (International Energy Agency). 2012. *World Energy Outlook*. http://www. worldenergyoutlook.org/

Khodeli, I., ed. 2009. *From Curse to Blessing? Using Natural Resources to Fuel Sustainable Development*. Hoboken: Wiley-Blackwell.

Klare, M. T. 2012. "The good, the bad, and the really, truly ugly." TomDispatch. http://www.tomdispatch.com/blog/175621/

Kraushaar J., and R. Ristinen. 1993. *Energy and Problems of a Technical Society*. Hoboken: Wiley.

MacKenzie, K. 2010. "How the majors see 'business as usual' on oil and climate." Financial Times. http://blogs.ft.com/energy-source/2010/03/29/how-the-majors-see-business-as-usual-on-oil-and-climate

Mystic Seaport Museum of America and the Sea. 2012. "The Charles W. Morgan: Preserving the Last Wooden Whaleship in the World." http://www.mysticseaport.org/index.cfm?fuseaction=home.viewPage&page_id=2103ED05-65B8-D398-7609445B7A947310

Schollnberger, W. E. 2006. From scarcity to plenty. Who shapes the future mix of primary energy? What might it be? *Oil and Gas European Magazine* 32: 8–20.

Shell (Royal Dutch Shell). 2011. *Signals and Signposts*. http://www.shell.com/global/future-energy/scenarios/signals-signposts.html

Strzelecki, M. 2012. "Coal to approach oil as top energy source by 2017, IEA says." Bloomberg. http://www.bloomberg.com/news/2012-12-18/coal-demand-forecast-to-rise-2-6-a-year-through-2017-iea-says.html

UN (United Nations). 2012. International Year of Sustainable Energy for All. http://www.un.org/en/events/sustainableenergyforall/

Index

CPSIA information can be obtained at www.ICGtesting.com
Printed in the USA
BVOW01*0246071013

332647BV00003B/4/P